Northeastern America

MANAGING EDITORS
Amy Bauman
Barbara J. Behm

CONTENT EDITORS
Amanda Barrickman
James I. Clark
Patricia Lantier
Charles P. Milne, Jr.
Katherine C. Noonan
Christine Snyder
Gary Turbak
William M. Vogt
Denise A. Wenger
Harold L. Willis
John Wolf

ASSISTANT EDITORS
Ann Angel
Michelle Dambeck
Barbara Murray
Renee Prink
Andrea J. Schneider

INDEXER
James I. Clark

ART/PRODUCTION
Suzanne Beck, Art Director
Andrew Rupniewski, Production Manager
Eileen Rickey, Typesetter

Copyright © 1992 Steck-Vaughn Company

Copyright © 1989 Raintree Publishers Limited Partnership for the English language edition.

Original text, photographs and illustrations copyright © 1985 Edizioni Vinicio de Lorentiis/Debate-Itaca.

All rights reserved. No part of the material protected by this copyright may be reproduced or utilized in any form by any means, electronic or mechanical, including photocopying, recording, or by any information storage and retrieval system, without permission in writing from the copyright owner. Requests for permission to make copies of any part of the work should be mailed to: Copyright Permissions, Steck-Vaughn Company, P.O. Box 26015, Austin, TX 78755. Printed in the United States of America.

Library of Congress Number: 88-18337

2 3 4 5 6 7 8 9 0   97 96 95 94 93 92

Library of Congress Cataloging-in-Publication Data

Wingfield, John C., 1948-
  [America settentrionale. English]
  Northeastern America / John C. Wingfield.

  — (World nature encyclopedia)
  Translation of: America settentrionale.
  Includes index.
  Summary: Discusses the plant and animal life of the northeastern area of the North American continent and its interaction with the environment.
  1. Ecology—Northeastern States—Juvenile literature.
2. Biotic communities—Northeastern States—Juvenile literature. [1. Ecology—Northeastern States. 2. Ecology—North America. 3. Northeastern States. 4. North America.]
I. Title. II. Series: Natura nel mondo. English.
QH104.5.N58W5613   1988   574.5′.264′0974—dc19   88-18421
ISBN 0-8172-3325-3

WORLD NATURE ENCYCLOPEDIA

Northeastern America

John C. Wingfield

Laramie Junior High
1355 N 22nd
Laramie, WY 82070

RAINTREE
STECK-VAUGHN
LIBRARY

Austin, Texas

CONTENTS

6 INTRODUCTION

9 A COMPLEX TOPOGRAPHY
The Ancient Mountains, 9. The Northern Regions, 10. The Great Lakes, 13. The Plains, 14. The Coast, 16. Acid Rain, 20.

23 THE GREAT LAKES
An Environment in Danger, 23. Fish of the Great Lakes, 25. Freshwater Coasts, 31. Winged Dwellers of the Swamps, 33.

37 THE FORESTS OF THE NORTHEAST
Deciduous Forests, 37. The Awakening of the Amphibians, 40. Reptiles, 46. Fish, 50. Small Fauna, 52.

57 BIRDS
The Return of the Migratory Birds, 57. Nonmigratory Birds, 58. A Myriad of Colors and Sounds, 61.

65 MAMMALS OF THE FOREST
Forest Dwellers, Large and Small, 65. Woodchucks and Other Rodents, 70. Underground Mammals, 73. The Gray Fox, 75.

77 SUMMER IN THE FORESTS
The Noisy Insect Community, 77. The Bane of the Forests, 79. Autumn Colors, 81.

83 WOODS IN THE NORTHERN REGIONS
The Trees, 83. The Birds, 84. The Mammals, 85. The Predators, 89.

91 WINTER
Life Under the Snow, 91. Adaptation to Life in the Snow, 93. Winter in the Deciduous Forest, 96.

99 THE ATLANTIC COAST
Atlantic Whales, 99. Colonies of Seabirds, 101. Sandy Beaches and Brackish Swamps, 103. Other Peculiar Animals, 104.

107 GUIDE TO AREAS OF NATURAL INTEREST
Canada, 107. United States, 112.

121 GLOSSARY

124 INDEX

INTRODUCTION

The eastern regions of North America have undergone more change and settlement than any other region on the continent. Pioneers and explorers first came to the region about three hundred years ago. Here they found wild territories, fringed by steep, deeply-indented coasts, sandy beaches, brackish swamps, and estuaries. Tides, dangerous coastlines, and sand shoals were difficult tests for even the most skilled sailors. Large rivers, such as the Hudson River, could be sailed upstream for over 90 miles (150 kilometers) with large boats. They served as gateways to vast inland territories. To the north, French explorers, called "voyageurs," sailed up the Saint Lawrence River all the way to the Great Lakes. This large waterway is one of the widest networks of fresh water in the world. The explorers used it as a corridor across the continent.

Eventually the explorers reached the Great Plains region. They were taken aback by this vast, flat grassland. Although it was treeless, the area was covered with flowers and teeming with wildlife. It was beyond every limit of imagination.

Of course, North America did not stay untouched for long. As word of this incredible environment spread, settlers began to pour in from Europe. Farms and villages spread inland. The deep, rich grassland soil, ranking with the finest soils in the world, was cultivated without pause. The forests were cut down and the grasslands were plowed.

During the Industrial Revolution, wide stretches of forest were even further damaged in the Northwest, and new towns appeared everywhere. All of these new towns were connected to major Midwest cities by a tight network of roads and highways. This became the industrial heart of the world's most important economy. The complexity and prosperity of this development is perhaps beyond compare. But because of it, the environment has suffered serious damage. The grassland has disappeared except for a few tiny relics. Rivers and swamps in many regions are now polluted. Even the Great Lakes—some as large as inland seas—have suffered from the changes. The fumes from smokestacks surrounding them have caused acid rain. This rain has killed all life-forms in many lakes.

But in Northeast America, the forests are returning. In some places, they once again cover their original territories. Some of this land is reclaimed farmland from New York state and New England. These farms were abandoned when the soil was no longer productive. Today, these areas are again covered with forests.

A COMPLEX TOPOGRAPHY

Most of the rocks of Northeast America are very old. Many are over a billion years old. The most ancient mountains form a chain which runs in a southwest to northeast direction. These are the Appalachian Mountains, which extend as far north as the New England states, New York State, and the coastal provinces of Canada. The region has also been influenced by the most recent glacial period. This occurred less than fifteen thousand years ago. The glacial periods have left their mark on the landscape. The entire area, thus, is a mosaic of ancient rocks and more recent modifications, caused by the Pleistocene glaciations.

The Ancient Mountains

The mountains of the eastern regions are called the Appalachian Mountains. They are the remains of a mountain chain that used to be much higher. These mountains are the result of a lifting process, erosion, further lifting, folding, and more erosion. All of this occurred over a period of a billion years. More recently, they were covered and smoothed by the glaciers which descended from the northernmost regions. These mountains do not have the wild majesty of the western chains. Only a few peaks are higher than 5,000 feet (1,500 meters). Many are rounded, with almost flat tops, due to the action of ice. All of these mountains are covered by a thick layer of deciduous forests. Deciduous trees are those that shed their leaves at a certain time each year. Conifer, or cone-bearing trees, dominate at higher altitudes. Below many of the vegetation-covered cliffs, canyons often open wide. There, rock landslides are frequent.

The Green Mountains in Vermont, however, originated from layers of sediment. This sediment accumulated at the bottom of an ancient sea over a billion years ago. These sediments then underwent a process of folding and twisting. They were then lifted up to form huge mountains, eroded, and eventually became the bottom of a new sea. All of this happened about 525 million years ago. Eventually, through plate tectonics, another mountain chain arose and was in turn eroded. Plate tectonics is the theory which proposes that the earth's outer shell, or lithosphere, is made up of numerous rigid sections known as "plates." According to the theory, these plates are slowly and constantly moving, interacting with each other in various ways. The pressure from the moving plates produces earthquakes, volcanoes, and mountains.

Preceding pages: A never-ending terrain and large wild expanses used to be the landscape which the first settlers to Northeast America saw. Even today, such regions can be found beyond heavily-populated areas, as this picture taken in Quebec, Canada, shows. Here, wide grassland alternates with thick conifer forests.

Opposite: Huge coniferous forests—either pure, as shown here, or mixed with deciduous trees—cover most of the northeastern regions of America. The forest starts from 984 to 1,312 feet (300 to 400 m) in elevation. These regions also are crossed by a great many rivers and streams, and are dotted with hundreds of small lakes.

A thick coniferous forest rims a stretch of grassland. In the foreground, among the grasses, is a marshy area. The varied features of the natural environments found in North America are the result of a succession of complex geological changes. These include the rise of huge mountains and their progressive erosion. The erosion process, which reached its peak during the last glacial period, caused massive surface changes.

Finally, during the glacial periods, huge glaciers flattened mountains, cut deep valleys, and carried masses of rubble down from the north. Many of these glaciers were thousands of feet thick. In the last million years, glaciers have moved forward and retreated four times.

When the glaciers retreated, they left behind rocks and soil which had a northern origin. Formations known as "drumlins" and "moraines" resulted. Drumlins are long, oval hills of glacial rubble. The steep side of the hill faces the direction from which the ice advanced; the gentler slope points in the direction the ice moved. Moraines are simply layers or ridges of rubble which form at the sides or ends of a glacier. These heaped materials obstructed valleys which then filled with water, creating lakes. When the glaciers melted, the sea level rose, and the ocean invaded the Saint Lawrence and Champlain valleys.

The Northern Regions

North of these mountains, and north and west of the Great Lakes, is a wide area covered by conifer forests. Some

Below is a cross section of a valley, the way it looks after the retreat of a glacier. Traces of a glacier's presence, such as little lakes, moraines, and erratic boulders are often found in the northern regions of the United States and Canada. From such traces, scientists know that these regions were covered by these huge masses of ice fifteen thousand years ago during the Pleistocene epoch.

deciduous trees—mainly birch and aspen—are also mixed in these forests. This is a region of plains or rolling hills. In it, some of the oldest rocks on earth can be found. They date back to 1.7 billion years. Some Minnesota rocks are as old as 2.5 billion years. This is about half the age of the earth. These rocks originally formed from lava intrusions pushing through cracks on the bottom of an inland sea.

Rocks found more to the north, on the Canadian Shield, are probably even older. The Canadian Shield is a vast, horseshoe-shaped region that covers almost half of Canada. It stretches in an arched pattern for over 1,860 miles (3,000 kilometers) from the Atlantic Ocean to the Arctic. It was once the bottom of a primeval sea. This sea, however, was invaded by ice and snow thousands of feet deep. This glacier gradually moved south, invading huge regions, smoothing and buckling the area under its great weight. As more and more ice accumulated, its mass became heavier and heavier. Eventually, the ground began to sink. Thus the Hudson Bay and the Great Lakes depressions were formed.

The abrasive action of the glacier freed the rock surface of all soil and rocky rubble. When the ice receded, the depressions in the ground filled with water. This created the hundreds of lakes which even today lace the northern

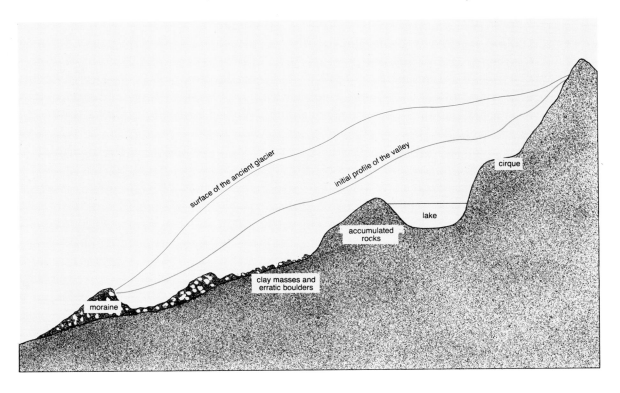

This unusual and enchanting photo shows Niagara Falls in winter. These falls, which connect Lake Erie with Lake Ontario, are an important power source. In order not to spoil the impressive natural phenomenon, the falls are only partially developed.

regions. Large boulders, carried by the glaciers and abandoned along the way, are still a typical feature of the landscape. Violent winds, which found no mountains to block their path, swept through the flat and barren territories. These winds carried most of the soil westward and southward, depositing it in the Great Plains area. There these materials added to the already deep, rich soil, forming what is considered to be one of the most fertile farming regions in the world.

The Great Lakes

The first explorers were frightened by the huge stretches of fresh water which they found inland on the American continent. Heavy storms can sweep across the lakes, creating large waves which break on the shores like ocean waves. This series of five lakes contains one-fifth of the freshwater surface of the earth. The lakes are relatively shallow. In fact, they are seldom deeper than 980 feet (300 m). They were created by glaciers between 10,000 and 15,000 years ago.

The Saint Lawrence River is the only major river flowing out of the Great Lakes. It flows toward the Atlantic Ocean and is still a major waterway for industrial transport today. In the past, these commercial routes were blocked by the Niagara Falls. This magnificently beautiful waterfall is formed by the collected waters of Lake Erie. The water falls in two streams into Lake Ontario. The larger stream falls on the Canadian side of the river, forming Horseshoe Falls. The smaller stream falls over the eastern shore and forms the American Falls. Together the two streams create a spectacular show of water power. Today, boats can easily move back and forth between the two lakes, thanks to the Welland Canal. The Canadian government built this artificial channel to be used as a shipping route. Niagara Falls, though, continues roaring nearby, unaware of the innovations of humans.

Lake Erie is the shallowest of the Great Lakes. For this reason, it can freeze completely in winter. The other lakes, on the contrary, are much deeper and wider. The action of wind and waves prevents ice from forming over their entire surfaces. Some ice can form along the shores, often in thick layers.

Immediately to the east of each lake is an area called the "snow belt." Cold winter winds which blow from inland carry warmth and moisture away from the lakes. These winds blow mainly toward the south and east. When they reach the land, they cool off again, and drop most of this moisture in the form of snow. As a result, snowfall over areas included in this belt is often heavy. In fact, it can be two or three times greater than the average amount over the rest of the region. On the other hand, these bodies of water warm up during the summer, and in autumn cool off more slowly than the ground. Like the ocean, they store warmth. This keeps the temperature along the coasts higher than average and delays the arrival of the first frost.

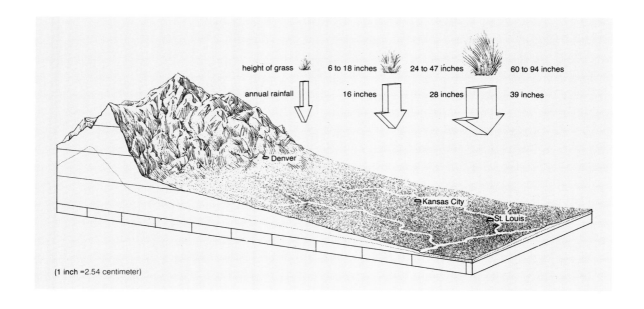

Between fifteen and sixty million years ago, the Rocky Mountains rose, breaking the monotony of North America's flat expanses. Material washed down from the eastern slopes of the chain helped to form the wide arch of the continent's central plains. Grasses in this area differ from region to region, depending on the amount of rainfall they receive. The western plains, in fact, are affected by the Rocky Mountains, which block moist air masses coming from the Pacific Ocean. Consequently, these plains receive very little rainfall. The grass remains short, growing in a more scattered pattern here than in the eastern plains. The eastern plains receive rainfall from moist air masses coming from both the Gulf of Mexico and from the Atlantic Ocean.

The Plains

West of the Great Lakes is one of the largest plains in the world. It stretches for about 3,000 miles (4,827 km) from northern Mexico to the subarctic regions of Canada. This is the largest plain in the central United States. It is shaped like a wide arch covered with grassland, narrowing into a point at its ends. The western plains are wetter than the eastern plains, as they receive a considerable amount of rainfall. Due to this climate, some grasses can grow to be over 6 feet (2 m) tall. In spring and summer, there are spectacular blooms, stretching as far as the eye can see.

The first explorers were amazed at the marvelous sight of the blossoming western grasslands. Hundreds of species of plants were found in these plains. The mix of these plants created a kaleidoscope of color. The sight of the grasslands, covered with tall grasses and flowers, must have been marvelous. But today, it no longer exists.

Most of the rain falls from April to July, in heavy thunderstorms that sweep the plains. Humid masses of air, coming from the Gulf of Mexico, move north, finding no barriers along the way. However, these air masses do meet with currents of cold and dry air which come down from central Canada. As a result, masses of hot, humid air rapidly rise over masses of heavier cold air. This causes the formation of storm clouds which stretch along the entire horizon, sometimes almost obscuring the sun. During these storms, rain-

The blossoming of the American grasslands is an incredibly beautiful show. The grasslands are known all over the world for an abundance and variety of colorful flowers. The communities of insects and spiders, which spend their entire lives hidden in the grass, are just as varied.

fall can average 8 to 31 inches (200 to 800 millimeters) in a few minutes. Luckily, these storms do not last long. But as they ravage the area, they are accompanied by lightning, thunder, and high winds. At times, hail as large as marbles can develop. Falling at high speeds from thousands of feet above, the hail can seriously damage the vegetation.

Undoubtedly, the most destructive forces of these storms are the legendary tornadoes. The tornadoes sometimes form when masses of hot and humid air coming from the south meet with cold, dry air from the north. The resulting movements are so violent that twisting currents join to form a sort of funnel. When this funnel touches the ground, a large-sized tornado is formed. These tornadoes are among the most destructive and powerful storms on earth. They are often accompanied by winds which blow at several hundred miles per hour. At ground level, the funnel moves much more slowly. Its speed ranges from 3 to 25 miles (5 to

40 km) per hour. The power of the whirling air sucks up anything in its path. Even cars can be lifted by the vortex. Often these are dropped back onto the ground hundreds of feet away. Houses, trees, and other large objects can be destroyed in a few seconds. The feeling of desolation and destruction which follows a sweeping tornado is hard to describe. Luckily, these heavy storms last only a few minutes. They cover rather short distances before dying off.

Summers on the plains are humid. The temperatures may rise as high as 90 to 100 degrees Fahrenheit (32 to 38 Celsius). On the other hand, winters are cold, with strong winds accumulating snow in large drifts. Then temperatures may drop to -22° F (-30° C) or lower. In spite of these extremes, the rich soil is very appealing to the farmer.

The Coast

The Atlantic coastline is deeply indented with bays, inlets resembling fiords, and huge estuaries. It is also known for its long, sandy beaches, coastal islands, and swamps. To the northeast, there is the island of Newfoundland. Newfoundland marks the north end of the chain of mountains and hills which forms the Appalachian Mountains.

Most of the northern Atlantic coast is rocky. It was formed by large volcanic intrusions into sedimentary rock. The sedimentary rocks are mainly sandstone, limestone, clay, and so on. In the past, glaciers eroded and reshaped these formations. Today, ocean waves keep this erosion process going. In New England, volcanic rocks, formed over one billion years ago, were submerged by various inland seas. The sediments on the bottom of these seas became solid, forming sandstone and limestone. Later, movement and pressure created by the earth's plates lifted and compressed these rocks. They underwent great changes, becoming marble, clay, and schist. Eventually, around 400 million years ago, faults and cracks opened up in this region. Weathering and erosion wore these old mountain chains down to their very foundations.

The Atlantic coastline is extremely varied. In some places, there are deep tidal bays. There the ocean rushes in and out twice a day, causing huge changes in the water depth. In some areas, such as in Cape Cod, the coarse sand is continually moved around by the ocean. Long stretches of sandy beaches extend for hundreds of miles southward. The central beaches of the Atlantic Ocean are typically accompanied by sandy shallows offshore. Calm lagoons,

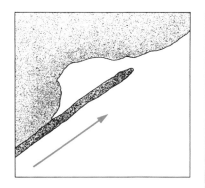

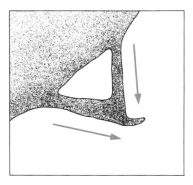

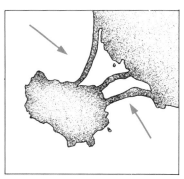

The Atlantic coast is particularly varied and interesting. The drawing shows how some of the most typical kinds of lagoons form. The arrows indicate the direction in which the sediment moves. *Top:* A coastal sandbar partially closes off a bay. *Center:* A jut of land, formed by the alternating movement of sediment, encloses a pool of water. *Bottom:* An island becomes connected to the coast by sediment tongues, which trap lagoons between them.

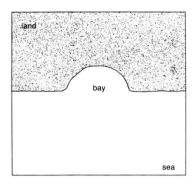

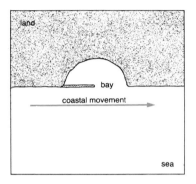

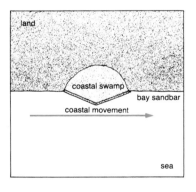

Another common habitat along the Atlantic coast is the coastal swamp. After a bay *(top)* has been partially closed by a sandbar, *(middle)*, the resulting lagoon can be completely cut off from the sea as sediments continue to accumulate. As this happens, water is no longer replenished inside the isolated lagoon. Soon, swampy vegetation is established. The lagoon thus becomes an excellent habitat for numerous species of water birds.

surrounded by swamps, separate the shallows from the coast. Elsewhere, tidal islands were created when the sea level rose due to the melting glaciers. Wide but shallow bays, such as Chesapeake Bay, are also typical of the Atlantic coast.

But how did such a varied coastline originate? For one thing, ocean waves can erode the rocky coast. This is especially true of the sedimentary rocks, which are softer. When particles and fragments of rock detach, they fall into the water and add to the waves' erosive action. Because of the water's continual movement, tiny particles such as clay do not sink. But in deeper waters, there is less turbulence and particles will settle to the bottom. There, they form layers of silt. Near the beaches, gravel and larger fragments deposit first, creating typical beaches. Where the action of the water is more powerful, pebbly beaches form. On the other hand, in the bays and estuaries, the water is calmer. There, tiny particles will settle to the bottom and create silt banks. Here, a swampy environment will eventually develop.

If the waves break directly onto the beach, sand is carried away with each wave and then returned. This leaves the sand more or less in the same place. Sometimes, however, waves break diagonally onto the coast. In this case, the waves carry sand in the same direction as their motion. If the shoreline is curved, sand will be deposited in deeper waters. Eventually, this creates a tongue of land at right angles to the coast. As the sand accumulates, a projection forms little by little. This may eventually develop into a peninsula.

Wind also plays a part in developing the coastlines. Blowing along the coast, the wind heaps up sand in certain spots. This soon creates sand dunes on which hardy plants may take root. The winds which blow at low tide are most effective for this. At that time, the shores are more exposed to their action. The surface layers of sand then blow into the dunes and are trapped there. This further enlarges the surface of the small peninsula. Vegetation makes the dunes more stable. They will stand firm even when violent storms bring waves that flood the peninsula, separating it from the coast. In this way, an island is born.

When the level of the sea rose, the coastal plains were flooded. Silt and mud were deposited at the bottom of the quiet backwaters between the coast and the islands. Large brackish swamps were formed. Many large rivers flow from inland into the Atlantic Ocean, carrying silt with the flow.

When they reach the ocean, the smallest particles form deposits. They create wide, muddy tidal swamps. Plant communities are soon established in these environments. They secure the mud with their roots and trap more sediment at high tide. Gradually, mud and vegetation pile up. Eventually, the tidal swamp expands beyond the coastline. In this way, a brackish swamp is formed. This kind of swamp can also develop when a small promontory or coastal island forms. Incredible numbers of organisms grow and thrive in these environments.

Perhaps the most spectacular peninsula is Cape Cod in Massachusetts. As the ice masses receded toward the end of the most recent glaciation, they left behind huge quantities of soil and rocks. All of this material was dumped on the continental shelf of New England. At the time, this was a

A group of American egrets wades in quiet waters. These and other very beautiful birds are a frequent and pleasant surprise to visitors in the rich environments of the North American wetlands.

18

coastal plain. When the ice melted, the sea level rose. This plain was flooded, and waves began to erode its rim. Cape Cod started out as a small promontory on the central part of a moraine. Sand dunes rose to a height of 147 feet (45 m), were flattened by storms, and were then reformed. This process occurred over and over.

The coastal moraine is many feet tall. In some spots, it is as tall as 180 feet (55 m). It is easily gouged by heavy storms, especially by those moving along the coast. Violent waves erode the shores at the rate of about 3 feet (1 m) a year. Ledges form and then collapse under the action of waves. The waves break them into small fragments. The larger boulders are usually scattered along the beaches. These will eventually be eroded by the action of waves and tides. The peninsula itself thus provides the material for the for-

The drawing below shows the areas from which acid rain's main polluting agents come *(top)*. It also shows the regions most affected by acid rain *(bottom)*. The areas most heavily affected by this phenomenon are eastern Canada and the northeastern United States. The pollutants originate mainly in the Midwest, and are carried to the affected areas by air currents and clouds *(red arrows, top)*.

mation of the beaches. Sand dunes, as you may remember, can be highly mobile. But they are usually stabilized by vegetation. Small local fires, though, sometimes destroy the plant cover. The dunes can thus move again, pushed by winds blowing from the sea. As they move, they cover everything in their path.

Acid Rain

Most of the northern and northeastern regions are affected by the phenomenon known as acid rain. Lakes and

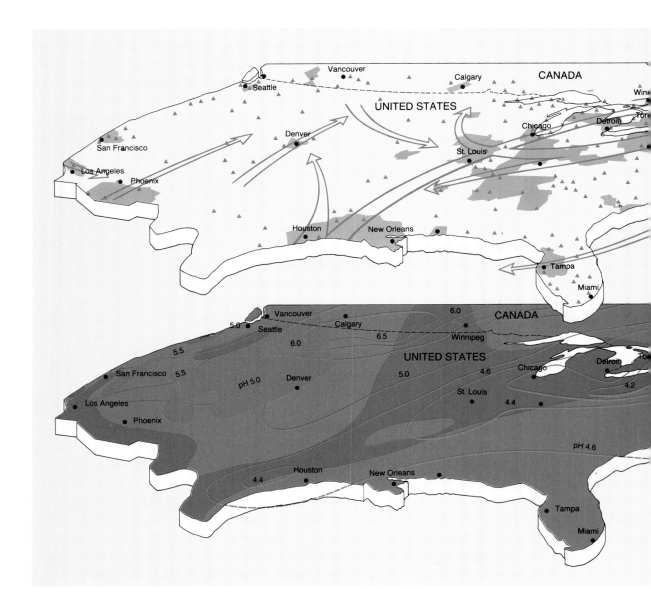

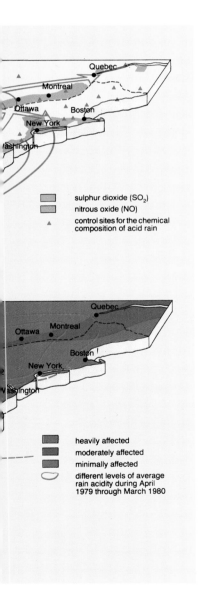

sulphur dioxide (SO$_2$)
nitrous oxide (NO)
control sites for the chemical composition of acid rain

heavily affected
moderately affected
minimally affected
different levels of average rain acidity during April 1979 through March 1980

mountain streams are polluted by rain and snow which contain toxic substances, mostly acids. Pollution caused by burning coal, mineral oils, and natural gas is released into the atmosphere in the form of sulphur and nitrous oxides. These gases react with hydrogen and water. They form sulfuric acid and nitric acid, two highly corrosive substances. This pollution is thought to originate in the smokestacks of the large industrial plants in the Midwest. Ironically enough, smokestacks are meant to help blow away toxic fumes. In effect, however, they simply spread the polluting substances over wider areas.

Acid rain does less damage to cultivated fields and forests than to lakes and rivers. Some lakes and rivers have limestone formations at the bottom. These alkaline rocks tend to neutralize the effect of acid rain. Eventually, though, carbon monoxide forms. As this happens, the alkalinity of the water decreases, and this "buffer" effect no longer works.

Some lakes and rivers do not have alkaline rocks at the bottom. In these, the water's acidity level climbs dangerously high. Moreover, other substances, like aluminum salts, dissolve in the water, polluting the lakes even further. Mountain lakes are the most seriously affected, since rain on the mountains immediately absorbs most of the acids from the atmosphere.

When water freezes into snow or sleet, ice crystals separate from the other molecules. The acids concentrate on the outside of the crystals. Acids then fall to the ground with the ice crystals. The process of melting and refreezing which occurs in the snow causes a further concentration of acids on the outer layers of snow. In spring, when the snow finally melts, the acids tend to wash away first. The lakes and streams which collect these waters undergo a sudden increase in acidity. Such acidity kills fish.

At the very least, acid rain can also destroy the balance of nature in an area. In some cases, high acidity levels have killed invertebrate organisms. Invertebrates are animals without backbones. This loss interferes with the normal food chain, where small prey are eaten by predators which in turn are prey for larger predators and so on. As you can see, acid rain is a real threat to life, both directly and indirectly.

THE GREAT LAKES

This huge system of freshwater lakes is sometimes called the fourth American coast. The entire five-lake system extends for 745 miles (1,200 km) in an east-west direction, and for 500 miles (800 km) in a north-south direction. The water surface is one-fifth of the surface of all freshwater bodies in the world. Of the five, Lake Superior is the largest. It is 370 miles (600 km) long and 186 miles (300 km) wide. The Great Lakes are so large that they create a specific climate affecting the surrounding regions. The storms which originate over them form waves that break on the shores like ocean waves. The Great Lakes are not very deep. The average depth is between 328 and 656 feet (100 m to 200 m). In a very few places, however, it may be as much as 1,200 feet (370 m) deep.

An Environment in Danger

The coasts of the Great Lakes are varied, with rocky boulders, pebbly beaches, and even dunes. The lakes were once surrounded by thick forests, but in many places, these forests have completely disappeared. Often they were cut down for agricultural purposes. Traces of the lakes' primary forests still survive in the north. This mass removal of forests is known as "deforestation." This deforestation and an increase in farming activities have changed the drainage pattern for waters flowing into the lakes.

The shallower lakes are most seriously affected by pollution from farmland drainage waters. These waters—Lake Erie in particular—contain large amounts of phosphorus, nitrogen, and other substances. These pollutant chemicals affect the growth of the lake's phytoplankton. Phytoplankton are the plant members of the plankton community. Most of these plants are microscopic algae that use sunlight to produce food and energy. They are consumed by zooplankton, which are the tiny animals of the plankton community. Both of these groups provide food for small fish. The small fish are in turn eaten by larger predator fish. Some types of phytoplankton grow best in polluted water. The pollutants cause them to multiply rapidly, but an increased phytoplankton supply is not always a good thing. It can upset the natural balance of life in the water body. Some lakes, however, receive less polluted waters. Therefore, these lakes, which include Lake Superior, host less phytoplankton.

About a hundred years ago, lakes' phytoplankton population was mainly composed of diatoms. Diatoms are uni-

Opposite: Even in winter, Lake Superior offers a picturesque view. The Great Lakes, located between the United States and Canada, boasts the largest freshwater surface area in the world. It is formed by a series of large lakes which notably affect the climate of the region. Due to low winter temperatures, the smaller lakes can freeze completely. The major lakes, however, usually only freeze along the coast.

Even today, in spite of an alarming increase in industrial pollution in the past few years, the coasts and waters of the Great Lakes host numerous species of animals, especially water birds. These birds nest in large numbers along the shores. In the picture, a coot is spotted during its daily search for food.

cellular algae enclosed in "shells" strengthened by silica. These shells resist decay. At the time, other types of algae were less common. When polluting agents, such as phosphorus, started washing into the lakes, the other species of algae began to multiply rapidly. As a result, the amounts of nutrients available in the water decreased. This eliminated most of the silica, which is essential to diatom growth. As well, such rapid algal growth also affected the available nitrogen levels. This then triggered a decrease in most species of algae.

One type of algae, however, continued to thrive. This was the blue-green algae, which can take and use nitrogen from the air. These algae became dominant in most of the shallower lakes. Unfortunately, these algae are a poor food source for zooplankton and fish, especially for those living in deeper waters. Thus, the local fish populations were seriously affected. Their numbers have sharply decreased in recent times. But these are not the only cause of problems in the Great Lakes. Other chemicals are also pollutants, such as heavy metals which contaminate the flesh of fish. Excessive fishing has also taken its toll.

This series of drawings illustrates the degeneration and death of a river, as it collects non-treated, polluted drainage waters. The water is originally well-oxygenated *(top)* and hosts water plants, several species of fish, shrimp, insects, and mollusks. When polluting substances enter the water, decomposition bacteria multiply *(middle)*, consuming all of the oxygen in the water. Mosquito larvae, leeches, and worms take the place of plant and animal communities. As well, a layer of black, slimy, and odorous algae builds on the bottom. In the final stage *(bottom)*, all of the oxygen has been consumed. Only certain bacteria can now survive in the water.

Fish of The Great Lakes

Numerous species of fish live in the Great Lakes. Some are found only in a certain area; others are common to all major North American lakes. The Atlantic salmon lives in the ocean, but when it matures, it migrates. It makes its way up rivers and streams, where it eventually lays its eggs. Once this species was very common in Lake Ontario, which it reached through the Saint Lawrence River. The salmon cannot reach the other lakes because of Niagara Falls, which are an insurmountable obstacle.

Adult Atlantic salmon may grow 3 feet (1 m) long and weigh as much as 44 pounds (20 kilograms). At spawning time, they swim up river to smaller streams. There they lay their eggs in shallow waters. They dig a very simple nest, called a "redd," on the stream bottom and lay their eggs in the loose rubble. The young fish, called "fry" spend about one year in the rivers, then began to swim down toward the lakes. From there, they swim into the Saint Lawrence River and make their way to the ocean. During this journey young salmon, called "parrs," undergo a transformation. They take on a silver color and become sea fish. The fish will stay in the ocean for about two years, then migrate again into fresh waters to spawn. Unlike the Pacific salmon, some of which have been introduced in the Great Lakes for sport and commercial fishing, the adult Atlantic salmon does not always die after laying its eggs. It may then migrate two, three, or more times in its life.

The lake whitefish and the cisco are freshwater fish. They usually live in the deeper lake waters, but venture into shallow waters in autumn to reproduce. Both are a silver color, with a light greenish brown back. Fifty years ago, record catches of these fish weighed up to 22 pounds (10 kg). Today, these fish are much smaller in size.

The lake whitefish feeds on the lake bottom, hunting small fish and invertebrates. More rarely, it will even feed on zooplankton. The cisco, on the other hand, seems to feed mainly on plankton. In spring, the adult ciscos move from deep waters up to the surface, where they feed on fast-growing plankton and on small invertebrates. As time passes, though, and the water temperature starts rising, these fish gradually move back to deeper waters. There the temperature remains lower.

The lake trout, spikes, and freshwater burbot are the main predators of the various species of whitefish. Among them, the chub is today an endangered species. This fish

The Saint Lawrence River is an impressive waterway. It is the major outflow for most of the Great Lakes. Toward its mouth, at Terranova Island, is a wide fiord. The first French explorers and settlers penetrated inland in this region by sailing up the river, which is navigable for a long stretch. They sailed all the way to the Great Lakes, which must have looked like a new sea to their eyes. Many parts of the riverbanks are well preserved even today. This allows for the survival of various water communities, both animal and plant.

Pictured are some of the most typical species of fish found in the Great Lakes. They include (*top to bottom, left to right*): a gigantic muskie seizes the leg of a duck with a swift movement; an alewife; female and male Atlantic salmon, as they respectively lay and fertilize their eggs on the bottom; a freshwater burbot.

dwells in open waters, but it stays in deep water its entire life, feeding and reproducing at levels deeper than 197 feet (60 m). Once this species was common in Lake Superior and Lake Huron (the deepest of the five lakes). Today it survives only in the latter.

Another dweller of the depths of the Great Lakes is the fourhorn sculpin. This fish, only 2 to 3 inches (50 to 70 mm) long, is a relic of the glacial period. During the last glaciation, when the glaciers covered most of North America, some populations of this fish moved south from their original habitat. When the ice receded, the fish were isolated in the glacial lakes. Today, they usually live at a depth of about 200 feet (60 m). But in Lake Superior, they are found as far down as 1,214 feet (370 m). They feed mainly on small invertebrates and reproduce in late summer or autumn. The sculpin are a very important food source for larger fish.

One predator of the tiny sculpin is the lake trout. This predator can be 1.5 feet (50 centimeters) long and weigh up to 44 pounds (20 kg). It lives in all of the major North American lakes. In autumn, it reproduces in shallow waters, but for the rest of the year, it lives in deeper waters. It is often found at depths from 65 to 200 feet (20 to 60 m) in the Great Lakes, but it can dive down to 1,214 feet (370 m). These trout

A pike jumps out of the water, probably to catch an insect. This large fish is armed with special teeth. These teeth, located on the palate near the tongue, can bend backward when seizing prey in order to make swallowing easier. The teeth then straighten out again, preventing the prey from escaping.

are fierce predators, catching smaller fish, like the freshwater burbot. The burbot is a typical species of the northern lakes. This peculiar looking fish may be up to 1 foot (32 cm) long and weigh up to 17 pounds (8 kg). Its color varies from yellow to light or dark brown. This fish also reproduces in the shallow waters. Oddly enough, however, reproduction takes place in winter, below the frozen water surface. In summer it moves to deeper waters.

Among the predators, the most famous species is certainly the pike. The northern species grows up to 2.5 feet (75 cm) long and has gray-olive coloration. It is a very common fish and prefers rivers bordered by thick vegetation and bays in the lakes, rather than open and deep waters. This fish has a voracious appetite. It preys on fish, amphibians (animals that live on land and in water, but can't breathe underwater), birds, and even small mammals which venture into the water. Even more impressive than the pike, is the muskellunge. This fish is said to grow as long as 6 feet (185 cm) long, and weigh over 99 pounds (45 kg). This species is also very common in bays with thick vegetation, as well as

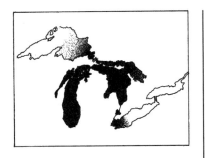

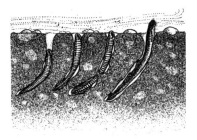

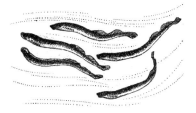

larger water bodies. It is known as "the great predator." Both the pike and the muskellunge spawn in early spring.

The fish population of the Great Lakes has undergone many changes. The native species have seen their numbers decrease. Other exotic fish have been introduced by humans or have actively invaded the lake system.

Among the latter is the alewife, a medium-size (0.5 foot- (15 cm) long plankton eater. This species is usually a dweller of salt waters, coming to the rivers and lakes to reproduce. In the last few years, the alewife has invaded the Great Lakes system, probably through the Erie Canal which connects the southern basin with the Great Lakes basin. Once in the lakes, the alewife population began to increase. It eventually took the place of other plankton-eating fish. At present, the alewives spend their entire life cycle in fresh water. They swim up tributary rivers and streams to spawn, then go back to the lakes for the rest of the year.

An even more serious problem is the invasion of the parasitic sea lamprey. This primitive fish grows about 3 feet (1 m) long and spends most of its life in the ocean. Unlike other vertebrates (animals with a backbone), the lamprey does not have jaws. Instead, it has a large, sucker-like mouth equipped with many sharp teeth. In the ocean, the lamprey is a parasite. Using its mouth and teeth, it attaches itself to other fish. It then scrapes a hole in the victim's skin and sucks its blood and body fluids. The fish is left with a typical round wound. Some fish, if frequently attacked, can even die. After several years of ocean life, the young lamprey matures and migrates into fresh waters. At this time, it stops eating. Mating and egg laying occur in small streams, where the male and female dig a shallow nest in the stream bottom. Afterwards, all adults die.

The sea lamprey has made many attempts at spreading into the lakes north of Lake Ontario. It probably bypassed the Niagara Falls through the Welland Channel (which connects Lake Ontario to Lake Erie), or perhaps it entered Lake Erie through the Erie Canal, like the alewife. In any case, today this fish is well adapted to the Great Lakes. Like the alewife, many of these lampreys now spend their entire lives in fresh waters. In the lakes, too, the lamprey is a parasite, preying on large fish like the lake trout. Its presence causes serious damage to fish populations, which already suffer from excessive fishing and pollution. It mounts the tributaries to reproduce, while during its parasitic phase it lives in the depths of the lakes.

Opposite: The top drawing shows the distribution of the sea lamprey in the Great Lakes during the 1960s. The drawings below it depict the lamprey's life cycle. Many useless attempts were made to trap and kill the lamprey with electrical barriers and traps as they migrated toward their spawning sites. Eventually, in 1958, scientists discovered a poison which can kill the lamprey at the larval stage. At this stage, they grow hidden in the silt. Using this poison, the pest was finally controlled. Before this, lamprey populations had been growing significantly since the turn of the century. They also had spread into all five lakes, seriously threatening many fish populations, especially lake trout.

Above: Water lilies frequently grow along the coast and in the swamps surrounding the Great Lakes. Their flowers add to the beauty of the natural landscape, which remains amazingly intact in many places.

Freshwater Coasts

The inlets, shallow bays, and swamps which surround the Great Lakes were once covered by a very characteristic vegetation. Today, these plant communities are threatened by water pollution. They are further disturbed by an increase in the activity of the water itself. This is caused by the growing number of motorboats which cruise the lakes. Much of the original flora has been lost, but some intact areas still remain.

Among the most common plants are the rushes and reeds. These plants stabilize the swamps, forming thick layers of vegetation. Some species, like the cattail and the bur reed, suffer damage from sudden increases in water level. An increase is often enough to completely wipe them out. Fortunately, their seeds survive well in mud and readily begin to grow, or germinate, after the water level decreases. For this reason, the swamps in the Great Lakes exhibit much more variety than the inland swamps, where the water level is more stable.

Other species adapt better to variations in water level.

A female mallard sits on its nest. Widespread in North America, this duck is one of the most common in the world. Numerous groups of mallards dwell in swamps. This duck is thought to be the ancestor of all domestic ducks. The male has a typical green head, with a white ring on its neck.

During mallard courtship rituals, the female invites the male, bending her head to one side *(top)*. The male answers the invitation with a sequence of typical postures. It shows off its bright plumage *(middle)*, and shows the ringlets formed by its tail feathers *(bottom)*.

Among these is the flowering rush, whose shape varies depending on water levels. In mud, it blossoms with pink flowers. When submerged, it does not blossom and only produces small, triangular leaves. Water lilies and the yellow spatterdock (*Nuphar advena*), produce longer leaf stems when the water rises. This allows their large leaves to float on or emerge from the surface of the water. Another species, the arrowhead, can grow as far from the beach as 30 feet (10 m), submerged in 24 to 59 inches (60 to 150 cm) of water. Its white blossoms are a common sight during the summer.

In some parts of the coast, low dunes alternate with swampy areas, as along the southern tip of Lake Michigan. On the western coast of Lake Michigan, "tidal" swamps are common in those areas where tributaries flow into the lake. The mouth of the Minch River in Wisconsin provides an example of this. These "tides" are caused by winds which blow lake waters into the mouth of a river. They are thus completely different from ocean tides. Lake tides can affect rivers several times during the day. The constant mingling of the river water, loaded with nutrients, with the lighter lake water, creates a very rich habitat.

Winged Dwellers of the Swamps

Among the birds which nest in the swamps are numerous ducks, such as the mallard, the blue-winged teal, and the black duck. The black duck used to be the most common permanent duck in North America. While it is still common today, its numbers are decreasing. The reasons for such a decrease are not completely clear, but the problem may, in part, stem from the crossbreeding of this species with the mallard. Oddly enough, the plumage of the male black duck is very different from that of a male mallard. The male mallard is basically gray, with a brown neck, a green head, and a narrow white band separating the two colorations. The male black duck, on the other hand, is a uniform dark brown, with a lighter head. Despite this, the two species are very similar, and crossbreeding is becoming more and more frequent.

All ducks nest very early in spring. The females lay ten or more eggs in nests built on the ground and wisely hidden. The ducklings are able to follow their mother to the water a few days after being born. In the middle of summer, the male duck molts. It then loses its bright plumage and takes on a duller coloration resembling the female's. This is called

Another common dweller of the northeastern swamps is the blue-winged teal. The teal shown here is in a swamp in Michigan. This is its favorite environment, between lake vegetation and open water surfaces. This species is easily recognized by the white, large crescent between its eyes and bill, and by the wide white patches on its wings.

cryptic plumage (for camouflage). At the same time, the males also molt their flight feathers, and for a certain period are incapable of flying. For this reason, they tend to gather in scattered groups, hiding in the vegetation of swamps and rivers, away from possible land predators. In autumn, they again take on a bright plumage.

Also, numerous species of the Rallidae family dwell in the swamps. The coot is often very abundant, and it is easily

recognized by its dark plumage, contrasting with a white bill and low, white forehead. It also has bright red eyes. The most spectacular species of this group is the king rail. This bird is rather large, often growing as much as 14 inches (35 cm) long. Despite its size, it is difficult to spot in the thick of the marshy vegetation because of its coloring. The king rail has a rusty-brown colored neck and chest, and its sides are striped with brown and white.

THE FORESTS OF THE NORTHEAST

The immense forests of the eastern part of North America once extended continuously from the eastern coast of Labrador to the Gulf of Mexico. They stretched inland all the way to the borders of the prairie, to Minnesota, Illinois, and farther to the south. The forests of the Northeast are mainly deciduous. Deciduous forests are composed of trees which shed their leaves in winter. But at elevations over 1,000 feet (300 m), conifer, or cone-bearing, trees dominate.

Deciduous Forests

The deciduous forests have been heavily cut down in the past. But in the last eighty years, many have grown back again in wide areas of Pennsylvania, New York, and New England. Many of the plant species which once lived in these areas are abundant again. These plants, together with new plants recently introduced by people, make up these new forests.

Among the species native to these forests are about twenty species of oak. Their seeds, the acorns, are a basic food source for many animals which dwell in the forests. Elderberries are another important food source. Elders are bushy plants which fill the undergrowth with their scented blossoms.

The larger trees in the forests, like the ash trees, tower up to about 80 feet (25 m). Another tree, the American basswood, can grow as much as 98 feet (30 m) tall. There are also many species of maples. One of these, the sugar maple, produces a sugary sap in late autumn that can be harvested to make maple syrup. The sap is also sought after by numerous animals, since it is a good source of energy. Maple trees grow to 100 feet (30 m) tall, and their trunks, with typical scaly and irregular bark, can be 47 inches (120 cm) in diameter.

The American sycamore tree is common along rivers and in other moist environments. This large tree can be up to 100 feet (30 m) tall, and has smooth, gray bark with scattered groups of brown scales. Its fruit, commonly called a buttonball, is formed of elongated seeds. When the fruit is ripe, it breaks loose. This allows the seeds to be blown away by the wind. American elms were also common at one time in the Northeast. They are still among the largest trees of the region. They can grow to be 115 feet (35 m) tall. Unfortunately, a fungus called Dutch elm disease was accidentally introduced from Europe. This disease has destroyed much of the elm population. Today, only a few specimens survive.

Opposite: The Adirondack Mountains, in New York, are within easy access of large urban areas which surround them. They are relatively low, covered by dense forests and mixed woods, and still offer natural environments in their original state, very rich in fauna and flora. These forests are a favorite place for nature excursions. Together with the more famous Vermont woods, they are well known for spectacular fall foliage colors.

The drawing shows some of the main tree species that make up the American deciduous forests. These trees are similar to those found in the forests of Europe and Asia. In the American Northeast, the main tree species are oaks, beech, maple, and shagbark hickory. In Europe and Asia, on the other hand, the main species are beech, chestnut, and basswood.

The American beech can also grow up to 100 feet (30 m) tall, and is very common. Its seeds are another important food source for animals, although they can be more or less abundant from year to year. The shagbark hickory and the walnut are also important fruit trees; they produce nuts in great numbers. The eastern red cedar is common along the edges of woods and in clearings. This is a conifer tree, which rarely grows more than 50 feet (15 m) tall. It grows in scattered groups surrounded by grass, forming the so-called red cedar grasslands.

Another species typical of the outskirts of woods is the staghorn sumac. Its densely hairy twigs grow vertically. They can be up to 36 feet (11 m) long but are usually much shorter. The sumac has large leaves that are dark green above, and lighter beneath. Its flowers are also green. In autumn, the tree's leaves turn scarlet, orange, and purple. At the same time, the flowers produce small seeds, each enclosed in a red, dry fruit, or drupe. In winter these fruits are an important source of nutrition for birds. They are thought to be rich in vitamin C.

Another species, which somewhat resembles the staghorn sumac, is the poison sumac. This plant, together with two similar species, the poison ivy and the poison oak, is common along the wooded ridges and clearings. The two latter species can grow up to 3 feet (1 m) tall, but the sumac

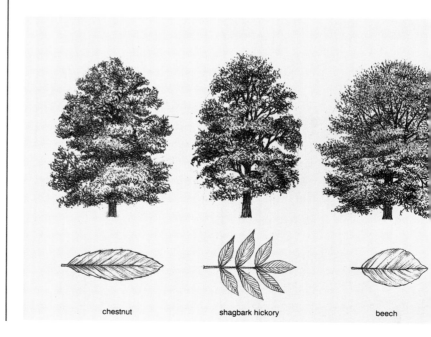

chestnut   shagbark hickory   beech

can be 29 feet (9 m) tall. Some of these plants are vines and climb on other bushes or trees, thus reaching considerable heights. All of these species produce a highly-toxic oil which causes serious skin irritations and sometimes even open sores. These plants are considered the terror of the Northeast. Some people have such a strong reaction to the poison that, after only one experience, they never go back into the woods during the summer. Oddly enough, most animals do not seem to be sensitive to this poison. Some birds even feed on the white berries of the poison sumac.

The skunk cabbage is one of the first wildflowers to emerge in the spring. This particularly interesting plant derives its name from the offensive odor which its leaves give off when crushed. This is a very common species in the northeastern woods, wherever the soil is moist. At first, a short, stout stem emerges from the ground, carrying tiny, tightly-grouped yellow or brown flowers. This stem, called a "spathe," is surrounded by a leaf-shaped sheath. These plants can grow in the snow or through the ice in early spring, pushing their way up through the frozen soil. Inside the sheath, the temperature is kept at 71°F (22°C). In order to produce this heat, the plant requires a great amount of oxygen. In early spring, the temperature inside the plant can be up to 54°F (30°C) higher than the soil and air temperature.

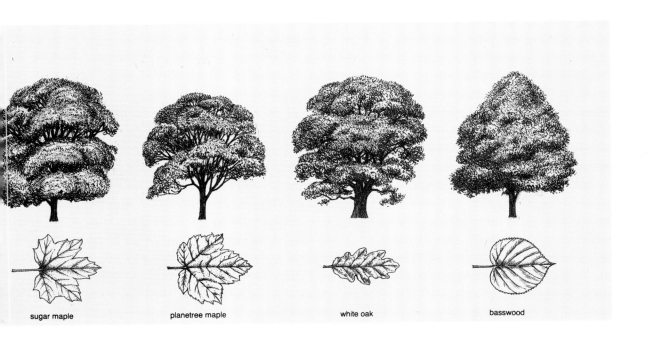

sugar maple   planetree maple   white oak   basswood

The wood frog belongs to a family in which the species differ widely in appearances and behavior patterns. These frogs are widespread throughout the world, except for southern regions such as South America and Australia.

Many invertebrates visit the skunk cabbage in order to warm up, and some spiders live inside it permanently, waiting for insects to come into its warm shelter. Its flowers are pollinated by bees and other insects which are attracted by its sweet smell.

The Awakening of the Amphibians

When spring arrives, plants and trees come to life, insects leave their hiding places, and many animals wake from their hibernation or arrive in migrations from the south. One of the first calls to be heard in the woods when spring comes is the love call of the wood frog. This frog is light or dark brown in color and is easily recognized by a dark line which runs from its nostrils to its eyes, then widens into a large stripe. This small frog, rarely longer than 2.7 inches (7 cm), almost always lives in moist woods throughout the region. It feeds on small invertebrates and worms. Its characteristic call, which rises from pools during mating season, is a succession of hoarse "quacks." It closely resembles a duck's call.

Large numbers of frogs appear together during mating

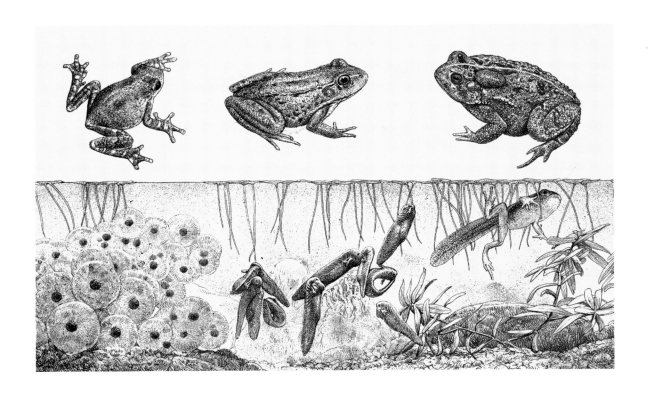

In the drawing are some of the most common amphibian species found in the wetlands of the American Northeast. These include from *left to right:* the tree frog, the green frog, and the American toad. At the bottom of the drawing, the main stages of a frog's growth are shown. Far left are the frog eggs, which are laid under the water in jelly-like clumps. In the center are a group of newborn tadpoles. Tadpoles are born completely legless and must breathe through their skin. Their gills will develop later. At the far right is the four-legged tadpole, swimming toward the surface. After the legs have developed, the tadpole's tail will gradually disappear, and lungs will take the place of gills. Soon metamorphosis into an adult frog will be completed.

season, when there is still ice on the pools. Then they may croak for a week without stopping. In the meantime, they lay masses of eggs, coated with a gelatinous substance. After the short mating period, the adults leave the pools and scatter away from the water.

Later in spring, another small frog appears and immediately makes its way to the pools to reproduce. This species has a very loud call, which it repeats every second. The call ends in a higher note, and sometimes with a sharp trill. The call has given this frog its name of spring peeper. In April and June, the call from a group of these frogs can be heard up to a mile away. The spring peepers are small and have a light rusty color. They may sometimes have a clear "X" marking on their backs. They prefer clear-water swamps, surrounded by shrubs and trees. After the mating season, they are difficult to spot.

The American toad also appears with the coming of spring and immediately looks for reproduction sites. This very common toad grows up to 3 inches (8 cm) long and is generally a brown or olive color. It has distinct, dark dots on its back, each of which contains one or two wart-like growths called "verrucae." In the swamps, the male's call is

The tree frog's call begins in a highly developed larynx, and is amplified by inflating throat sacs. These sacs can be single *(as the frog in the picture)* or double. They can be placed inside or outside of the throat, and their shapes and locations vary with the differing species.

a long, musical trill which lasts from six to thirty seconds. This call, together with that of the tree frog, is among the most familiar and pleasant sounds of the spring. The two species often live together, and their calls, from a distance, resemble ringing bells mingled with trills. The American toad croaks by using two protruding throat sacs, which it inflates by swallowing air. It mates from March to July, after which it retreats into the woods to moist areas, far from open water.

In late spring, when the temperature is higher, other amphibians appear. The two species of gray tree frogs are actually almost identical and very difficult to tell apart. They can be 2 inches (5 cm) long, and are gray, with lighter spots and brown markings. When resting on the bark of a tree, they are very well camouflaged and often their croaking is the only sign of their presence. The males are very territorial. Each one defends a small tree, indicating its position with a sequence of rhythmical and musical "barks." When they croak, they inflate vocal sacs located on the sides of their heads. These tree frogs are very active on warm, humid nights, and their call can be heard very far.

At this time, the calls of larger frogs, like the bullfrog and the green frog, rise from every pond. The call of the green frog resembles the sound of a banjo string, and it is very familiar on summer nights. The larger bullfrog can be 6 inches (15 cm) long and gives out a deep call, called a "jug" or "rum." Its mating season starts in June and goes on until July. Males emit their call mainly at night, with each having its own territory of 6 to 23 feet (2 to 7 m) in diameter. Frequent fights occur to defend the territories.

On these occasions, the males will give off deep, hoarse sounds. The females remain very secluded, getting close to males only when they are ready to mate. They seem to prefer the larger males, and each one lays thousands of eggs. The smaller males, defeated by larger rivals, are not able to lure the solitary females. For this reason, some adopt a rather tricky technique. They capture the females as they head to mate with the larger males. The females, at first, will frantically fight to get away. But they soon get tired and eventually lay their eggs. Despite this technique, the larger males still control the best territories, where tadpoles can safely develop. Smaller males, on the other hand, have access to females in less appealing areas. Even when they succeed in fertilizing some eggs, these may be eaten by predators, or the tadpoles will fail to survive for some other reason.

As well, a particular species of leech preys upon the frog eggs. It attacks clumps of eggs, and can destroy half of them before they hatch. Tadpoles are also easy prey, and are eaten by fish and invertebrates. Their death rate is so high that only one out of ten thousand eggs develops into a mature adult. For this reason, females must lay many thousands of eggs each mating season. After they have reached maturity, they live two to three more years. Of the thousands

A young bullfrog sits at the water's edge. The adults of this species can be 8 inches (20 cm) long, and weigh over 2 pounds (1 kg). At one time, this gigantic frog lived only in North America, but today it is found in various parts of the world.

of eggs they will lay in that time, only a few will reach full maturity.

In spring, other animals also appear in the woods and swamps. Newts and salamanders are two examples. Among the newts, the most famous species is the common newt. The adults of this species live in water—in pools, in lakes, and elsewhere. They are usually 2 to 4 inches (7 to 10 cm) long, and are a yellow-olive color, with vivid red spots along the sides. During mating season, the male develops a tall crest on its tail, which it uses during its mating parade.

Newts are usually active throughout the winter and have been seen swimming beneath the frozen surface. Their life cycle is complex. The tadpole turns into a land animal called an "eft," which is light red or orange, with typical darker red spots along its sides. These well-known red larvae abandon the water and live in the woods, often in full light. After about a year, the small efts reach maturity. Then they become darker but keep their red spots. At this time, they go back to the water, where they will spend the rest of their lives.

The red salamander is commonly found in shady places, close to ponds and streams throughout the eastern United States. The adult has beautiful coral red coloration, sometimes dotted by black spots. The newborn larvae, on the contrary, are a uniform dark yellow.

Another common species is the spotted salamander. It can be up to 8 inches (20 cm) long, and is a gray or bluish black color, with round yellow or orange spots in an irregular row along its sides. During the first warm rains, the hibernating animals wake up and come out from their shelters under logs, bushes, and so on. They move toward the ponds to reproduce. For the rest of the year, they live in the woods, and are very difficult to spot. In fact, they come out into the open only on rainy nights, to feed on invertebrates.

One of the most spectacular amphibians of the region is the hellbender. This is a giant salamander, up to 20 inches (50 cm) long. This animal is very strange looking. Its head is flattened, and fleshy folds of skin run along its sides. Its color varies from gray, to yellow-brown, to black. It lives in rivers and streams with running water, hiding under rocks. It feeds on small crustaceans, worms, and other aquatic invertebrates. Little is known about the hellbender because of its shy habits.

Almost all amphibians must go through a larval stage. During this stage, they are called tadpoles. Eventually they undergo a metamorphosis, turning into four-legged adults that lack external gills. However, some species never

common newt (larval stage)

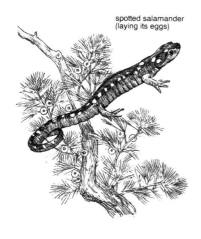
spotted salamander (laying its eggs)

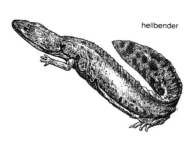
hellbender

mud puppy

undergo metamorphosis; they remain in an essentially larval stage. This phenomenon is called "neoteny." However, neotenic larvae do mature and are able to reproduce.

A common example of neoteny is the mud puppy. This animal lives in lakes, ponds, slow-flowing rivers, and streams throughout the region, except for the northernmost areas. It can be 12 inches (20 cm) long, and stays in the larval stage with clearly evident, feathery and brown-colored external gills. It has small legs with four toes. Its color varies from gray to rusty brown, and sometimes, indistinct dark markings run along its back. This salamander lives in water all its life and feeds on fish eggs, invertebrates, and mollusks. A mollusk is an invertebrate that has a soft body, usually covered by a hard shell.

Reptiles

Reptiles are also well represented in the region. Even so, only two species are widely distributed. They are the eastern swift and the five-lined skink. The swifts grow up to 7 inches (18 cm) long, and have a gray or brown coloration. Males have violet or greenish blue stripes along their sides underneath. They are often seen on fences and sometimes climb trees. They feed on insects and other invertebrates, and reproduce by laying eggs.

The five-lined skink can grow to almost 8 inches (20 cm) long. The young are brightly colored, with five white stripes on a black background, and a blue tail. When they grow to be adults, the stripes become less evident, the background color becomes duller, and the tail turns gray. The males, during mating season, take on an orange coloration around their mouths. These lizards live inside logs and rotting trees, in or near the woods. They prefer moist areas and are active during the day. They mainly feed on invertebrates.

Snakes are much more common in the region, represented by several species. Among them, the garter snake is the most common, since it lives in a great many environments. This is a medium-sized snake, up to 24 inches (60 cm) long. Its coloration can vary considerably; usually it has three yellowish stripes on a darker brown or black background. It lives in meadows, swamps, and woods, along rivers, and often even in gardens. When caught, it emits a strong musky odor from its anal glands as a means of defense. Its highly-varied diet is comprised of invertebrates, toads, salamanders, and even carrion. The female gives

Several species of garter snake are common to North America, and all have rather similar habits. This snake usually lives close to the water, in swamps, and grasslands. Mating occurs in March in the southern areas, and in April in areas more to the north. The young, which are born between July and August, are about 6 to 8 inches (15 to 20 cm) long. One female may give birth to as many as eighty young.

birth to small, living snakes.

Snakes spend the winter hibernating underground. In the northern areas, like Manitoba and Canada, these snakes hibernate in dense gatherings of ten thousand to fifteen thousand individuals. Hibernation sites are deep, underground limestone pits, where the temperature never falls below about 41°F (5°C). In spring, males emerge first and wait for females to begin courtship and mating rituals. One female may be courted by many males—sometimes fifty or more. The result is a tangle of snakes, sometimes called a "mating ball."

The water snake can be up to 3 feet (1 m) long in some areas. This snake also has highly variable coloration, but generally it is grayish, with alternating light and dark spots along its sides. It prefers water habitats, near ponds or rivers. It feeds on fish, frogs, toads, and salamanders. It is a skilled swimmer, and can dive under the water, looking for prey.

Poisonous snakes are not common in the Northeast, but the timber rattlesnake is found in many areas. This snake can be up to 4 feet (1.30 m) long. It is often yellowish

A water snake moves easily through the water plants which completely cover this pond's surface. This snake is another very common species and is widespread over the Northeast. It has strictly aquatic habits and is found from sea level to 4,921 feet (1,500 m) in altitude. It is completely harmless, and in some areas it is sometimes kept as a pet.

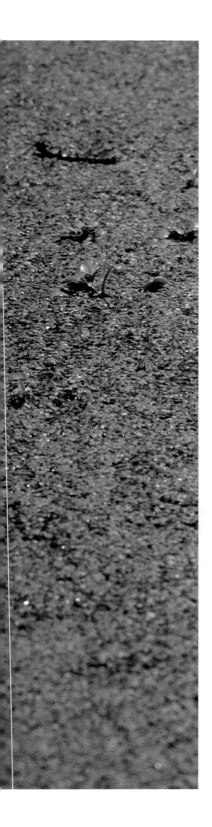

with darker bands. Sometimes, it is densely spotted with black and brown dots, so that the light background color appears darker. It is still rather common in the mountains of the Northeast, especially where there are rocky outcrops in the woods. Like all rattlers, this snake's tail is made up of loosely interlocking segments. When rapidly shaken, these segments produce a clear rattling sound. This signal warns intruders to clear out. Rattlesnakes prey on small mammals, which they seize with one rapid bite. The long, hollow poison fangs penetrate into the victim's body, and a very powerful poison is injected. The victim usually dies within a few minutes.

Swamps, ponds, rivers, and streams are also ideal habitats for many species of turtles. Some barely penetrate into the region from the south, and very few species live in the northern areas. These few species, though, are widely distributed and very common. The painted turtle is found in practically every lake and pond. Usually it is dark in color, with red markings on the rim of its shell and large yellow spots on its head. This turtle can grow to 6 inches (15 cm) long and prefers shallow waters with dense vegetation and muddy bottoms. It is an omnivorous animal. An omnivorous animal eats both plants and animals. Its diet includes water plants, insects, shrimp, mollusks, and other invertebrates.

Female turtles are usually larger than males and lay their eggs at different times. When it is time to lay their eggs, females leave the ponds and dig small holes in compact soil or sand. There they deposit their eggs and carefully cover them up. After several weeks, the small turtles emerge from the soil and crawl to the nearest pond. Their death rate is high due to raccoons and other predators which are always searching for a turtle egg treat. Despite predator problems, though, many young turtles survive. This species is abundant even today. These turtles are often seen on muddy banks or partially sunken logs, resting in the sun.

One of the largest turtles in the region is the common snapping turtle. This turtle can be up to 12 inches (30 cm) long and weigh 35 pounds (16 kg). It has a fierce look and a temper to match. Its head is large and its shell is a gray or light brown color. Its tail is long and serrated, or notched on the edge like a saw. These turtles are common in most permanent water bodies, but they do not usually rest in the sun like other species do. They are also omnivorous and feed on water vegetation, frogs, fish, small mammals, birds, and invertebrates. They stay hidden in the mud, allowing

The animal in the picture, commonly called a snapping turtle, is a member of a family of large freshwater turtles. The snapping turtle has a sturdy head, a powerful, hooked beak, and a long tail. These turtles live in most any body of water and feed on any kind of food. The females, which are much smaller than the males, lay large numbers of eggs.

only their heads to stick out. From this position, they can easily seize their prey. Out of the water, as they move from one pond to the next, or as the females dig holes to lay eggs, snapping turtles exhibit a fierce temper. If frightened, they strike swiftly, and their bite is very painful. To make them look even more terrifying, snapping turtles often have large numbers of leeches swollen with blood dangling from their skin. As you can imagine, this species may look repulsive at first sight. However, it is a very important element in the humid environment of the eastern regions.

Not all species of turtles live close to water. The wood tortoise, for example, lives in meadows and woods, away from the water, but it hibernates at the bottom of swamps and rivers. This species of tortoise can grow to be 8 inches (20 cm) long, and its shell bones slope upward. It has a brown color, with an orange head and legs.

Fish

The many rivers, pools, and lakes are an ideal habitat for large numbers of fish. The bluegill is common in most of

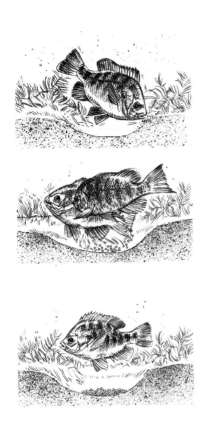

The bluegill lives in shallow waters that are rich in vegetation. It is well known for its peculiar reproductive behavior. The male "digs" a hole on the bottom with its ventral, or pelvic, fins *(top)*. The female then lays its eggs there *(middle)*. The male then watches over the eggs until they hatch *(bottom)*. In some areas, it is possible to find large groups of these peculiar nests. Each nest is attentively watched by the male which built it.

the larger water bodies. It has a green-brown back and lighter underpart. Numerous dark, vertical stripes start from a yellowish belly and run up its sides. The male's belly turns orange during the mating season. The adult bluegills measure 6 to 8 inches (17 to 20 cm) long. In summer, some males claim and defend a territory in shallow water. Often this territory is no more than 1.5 to 2 feet (45 to 60 cm) in diameter. Within this area, the male thoroughly clears a small, circular area of all vegetation and debris with swift movements of its tail and fins. The female enters the male's territory in July to lay its eggs. The male watches them until they hatch.

Recent research, though, has shown that not all males are territorial. These non-territorial males, as they are called, will often try to fertilize eggs in another's territory. The non-territorial males mature and are able to reproduce in two years. Territorial males, however, need six to seven years to fully mature. The non-territorial males are also much smaller than territorial males and manage to reach the nest as the female is laying its eggs. The defenders of the territory will try to chase the intruders away, but the non-territorial males are so small that they are often not spotted immediately. As these males grow older, they also grow bigger. As they do, they have a harder time entering another's territory without being attacked. They then resort to an alternative technique—they imitate the female's coloring. With this disguise, the non-territorial males manage to get close to a mating pair without being attacked by the male.

Smaller fish also abound in the waters of the American Northeast. Among them is the very common darter fish. This fish is rarely longer than 2 or 3 inches (5 to 8 cm). It reproduces in spring, and the male defends a territory typically located under a rock or in another secure spot. When a female enters a male's territory, the two fish mate belly up. The eggs are surrounded by a sticky coating with which they can be attached to the undersides of rocks. This habit can be advantageous in rivers with running waters, where eggs laid loosely on the bottom would easily be carried away.

The shiner fish are also common and widespread. They can be from 2 to 4 inches (5 to 10 cm) long, and they are brightly colored. These fish prefer streams and rushing brooks, and males are territorial. They lure females to lay their eggs on shallow pebbly bottoms of quieter spots in streams.

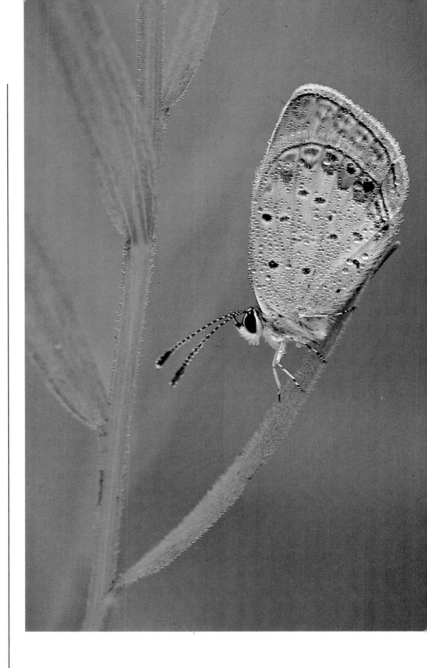

This elegant butterfly is a copper. Here it is photographed in the early morning, as the dew on its wings prove. The copper has a delicate pattern with eye-like markings on the lower side of its wings. This coloring contrasts with that of the upper side of the wings, which is often an even blue. Coppers live from sea level up to the highest elevations. They are mostly found in grasslands and clearings.

Small Fauna

In the Northeast, the spring season is one of the most spectacular periods of the entire year. It is a time when the plants grow and color the landscape, their fragrant blossoms freshly opened. Warm temperatures in March and April also bring large numbers of insects, even though the temperature may still drop and the weather may still vary.

Some of the insects which live in the northeastern woods, though, must find a way to deal with low tempera-

Among the moths belonging to the genus *Malacostoma*, a particularly well-known one is the *Malacostoma neustrium*, a tent caterpillar. This drawing shows an adult and typical masses of its eggs. The eggs, which look like a muff, are generally laid in the fork of a tree. During their growth, the larvae of this species can completely strip a tree of its leaves.

tures. Butterflies, for example, must raise their body temperature to 80°F (27°C) in order to fly. Actually, research has found the temperature in an insect's thorax, or mid-body section, to be much higher than this. Of the many species studied, the mean temperature was 95°F (35°C), with a minimum of 84°F (29°C) and a maximum of 104°F (40°C). Although they are cold-blooded, many insects manage to maintain an internal temperature similar to that of mammals and birds.

Butterflies are able to control their internal temperatures through special habits. Some butterflies, such as the monarch, the swallowtail, and the brushfooted butterflies, rest in the sun with their wings wide open. Some species, however, keep their wings closed. This exposes the lower body surface to the sun's rays. Other species keep their wings slightly open. This technique also allows the butterfly's body to receive most of the sun's warmth. The coppers and the garden whites both practice this method. Other species of butterflies rest their bodies on warm objects, absorbing heat from them. Thanks to all of these techniques, butterflies are able to fly very early in spring. Often they are flying before the temperatures get much higher than winter temperatures.

Many adult butterflies hibernate for the winter. When the air first warms up in spring, they can raise their body temperatures 14° to 18°F (8° to 10°C) above the air temperature through muscular contractions. These contractions can be compared to shivering in mammals. For this reason, these butterflies are among the first species to appear in the spring, and among the last to fly in autumn.

Some butterfly species display a quality known as polymorphism. Polymorphism is the ability of a species to assume different forms. It occurs in both plants and animals and is common among insects. In certain butterflies, different generations of the same species look different, depending on when the butterfly passed through its larval stage. Adults with dark lower wings develop from larvae which have grown in the early spring. Adults with light lower wings develop from larvae that grew in summer. The color variations help the larvae survive. The light color of the summer larvae keeps them from overheating. The dark color of the spring larvae helps them absorb the necessary warmth on cold spring days.

The moth larvae known as "tent caterpillars," live in colonies of up to three hundred individuals. Their name

In the northeastern regions, mosquito populations can be very numerous. Because of them, any trip into the woods can turn into physical agony because of the insects' constant biting. The mosquito larvae and nymphs are aquatic. They live wherever there is water, from puddles to lakes. From time to time, they come to the surface to breathe. In this picture, two mosquitoes reach the end of their metamorphosis. With appropriate movements, they come out of their nymphal sheaths, right onto the surface of the water.

comes from the large tent-shaped structures that they build. These tents are very evident on host trees, such as cherry trees and apple trees. An adult moth lays masses of eggs in autumn. A small larva immediately begins to develop inside each, but the eggs will not hatch until the spring. They are able to survive the winter months thanks to an accumulation of a chemical called "glycerol." This chemical acts as a thermal insulation, enabling the larvae to withstand temperatures of -22°F (-30°C) in January. Even if

spring is late, the larvae can survive thanks to nutritious substances stored in their eggs.

When tree buds begin to swell, the larvae break out of their eggs and start to feed on young, green leaves. The caterpillars move around, secreting a silky thread. This thread comes from silk glands located on the lower part of the caterpillars' heads. As the threads accumulate, they eventually form a very evident path. This path may aid the caterpillar as it moves along on the smooth bark.

The tents are built soon after the caterpillars hatch. They are continually added onto, growing larger and larger as the caterpillars grow. The weaving is done in groups, with a weaving session lasting up to an hour. During this time, the larvae spread over the structure's surface, laying a net of silk threads. The surface of the tent is very taut, and the inside has several layers and compartments. A tent is usually located on the open side of the tree and has a very evident entrance. Caterpillars are often seen near the entrance, leaving to feed on the tree. The tent absorbs the sun rays, and even on the coldest days of spring the inside temperature can reach 100°F (38°C). Heat gets inside the structure and concentrates there. The walls of the tent, though, are poorly insulated, and the heat is rapidly lost at night.

The larvae feed on the host tree and can completely strip it of all its leaves. They secrete a substance called "pheromone" to mark the paths which lead to the feeding areas. Other members of the colony can detect this substance. Following it, they will be able to find their way to the feeding area and back to their tent. Caterpillars also cooperate as they explore a tree. It has recently been proven that they somehow communicate among themselves, sharing information on good feeding spots. This behavior can be considered a result of so-called kin selection. Since all of the colony's members hatch from the same egg mass, they are all brothers and sisters.

Some species of highly-annoying spring insects also live in the forests. Numerous species of black flies appear in great numbers in the northeastern forests. These creatures are just 0.15 inches (4 mm) long, humpbacked, and black. The female black flies bite ferociously, sucking the blood they require for the development of their eggs. Swarms of these insects can be seen clouding around any animal or human walking in the woods. Visitors to some forests, especially in the northern regions, are told that woodcutters have gone mad because of these insects' bites.

BIRDS

Opposite: A Baltimore oriole stands at the entrance to its peculiar, sac-shaped nest. This bird belongs to the *Icteridae* family. Members of this family, which are typical in the Americas, are mostly tropical birds. Only a few species reach the northernmost regions of North America.

In the first days of spring, when snow and ice begin to thaw, the nonmigrating birds start their activities with new energy. Soon, the migratory birds return from their wintering areas in the south. The male red-wing blackbirds flood meadows and swamps, signaling their territories with loud songs and bold flashes of their red-tipped shoulders. The song sparrows arrive in great numbers; their lively refrains can be heard all over in the meadows, farmlands, and swamps.

The Return of the Migratory Birds

Larger birds, like the bronzed grackle, come in flocks. Initially, they feed around farms, then slowly scatter all over the region, forming pairs. The male of this species, seen from a distance, looks black. With a closer look, however, the bird's plumage takes on a bronze tint, especially in full light. The males also have long tails, which they stretch out in a characteristic fashion when they meet other males or when they court females.

Perhaps the most peculiar migratory bird is the bluebird. The male of this species has a bright blue back and tail, and a vivid rusty red chest. The female has a grayish back with a blue tint and a rusty colored chest. At 5.5 inches (14 cm) long, bluebirds are rather small. They build their nests inside tree holes and logs.

About twenty years ago, the populations of these birds decreased alarmingly. This decrease may have resulted from the excessive use of highly-toxic pesticides. Competition with other birds for good nesting spots might have been another reason. Competition with the starling, which was introduced from Europe, was especially intense. Recently though, the use of pesticides has been limited, and local groups of naturalists and conservationists have built artificial nests for the bluebirds. The nests have entrance holes which are large enough for a bluebird but too small for other species. This project has been very successful, and today the bluebirds are again common in the northeastern territories. The main competitor today is the tree swallow. This is a small swallow, with a black greenish iridescent back, contrasting with snow-white underparts. This swallow also nests in holes. Both species have probably taken advantage of the artificial nests.

One of the most interesting first arrivals is the American woodcock. This species lives only in moist woods and swamps. It is about 8 inches (20 cm) long, and its underparts

red-winged blackbird

bluebird

tree swallow

American woodcock

are a rusty color with a complex network of brown stripes. Its back is gray. This plumage, as mentioned earlier, is a good example of mimicry. A woodcock crouched on the leafy ground is practically invisible, and the bird knows how to use this camouflage. When people approach, it stays completely motionless. It will move only if it is just about to be trampled, and then it will swiftly fly away. This action is too rapid an event for a predator. The predator's shock allows the woodcock to escape. It will dart among the trees and sneak into the thick underbrush.

The woodcock's bill is almost one-third of its body's length. At the tip of the bill there are nerve endings, so the woodcock can sense worms or other invertebrates when it sticks its bill into the mud or wet soil. The woodcock also has rather large eyes, as might be expected in this mainly nocturnal animal. The eyes protrude from the head and allow for 360-degree sight. This ability allows the bird to spot an approaching predator, while it is busy looking for food on the forest floor.

In spring, the woodcocks give out one of the most familiar calls in the woods. The males begin their courtship parades on the forest floor or on small reliefs, giving off peculiar clucking sounds. They fan their tails and stretch their wings downward. From time to time, they fly in circles. The purpose of the male's parade and song is to define a territory. This display is also used to attract a mate.

Nonmigratory Birds

Not all birds are migratory. Some stay in the same region all winter. But these birds will also start singing and showing off during the first few days of spring, especially if the weather is warm and sunny. The blue jay is one of the most common birds of the Northeast. It is a large bird, about 10 inches (25 cm) long. It roams the woods and suburban gardens in scattered, noisy groups, giving out a great many sounds. The most common call is a screeching "kiar-kiar." The blue jay is an attractive bird, with a bluish back, tail, and crest. Its sides are dotted with black markings which extend under the bill, forming a narrow band. The blue jay's bright coloring, as with many birds, is not the result of pigment alone. Rather some tints are caused by the specialized structure of the feathers. These structures absorb certain wavelengths of light and reflect others toward the eye of an observer. In this case, the feathers reflect mainly the blue color. On gray, rainy days, when there is no direct

Perhaps one of the most familiar birds of North America is the blue jay. This bird has little fear of people, is noisy, and has elegant coloring. It often dwells in city parks and on the outskirts of towns. It may even nest close to houses, where it will even eat from bird feeders generally filled for birds with more "urban" habits. The crest on the blue jay's head gives it a very distinct appearance and may account for its popularity.

sunlight, the blue jays look markedly gray, black, or whitish.

Another common nonmigratory bird is the ruffed grouse, which has peculiar erectile feathers on its neck. This bird can be up to 14 inches (35 cm) long and has brown, rust, or sometimes gray stripes on its underparts. It feeds mainly on plant matter and often also on invertebrates. In spring, grouse are often seen perched on maple tree branches, pecking at their flowers.

From March to June, the male ruffed grouse engages in a peculiar parade. Perched on a log, the bird raises its neck feathers and fans its bright, rusty-colored tail in a perfect semicircle. At intervals, the bird makes a low-pitched drumming sound by beating the air with its wings. These sounds start at the rate of about one per second, speed up, and then slow down again. The males vigorously compete for the best logs on which to perform their drumming. Logs of a certain size, especially if hollow, probably tend to amplify the drumming sound. In the first days of spring, the female ruffed grouse visits numerous parading males and eventually chooses a mate. She then builds a nest and lays

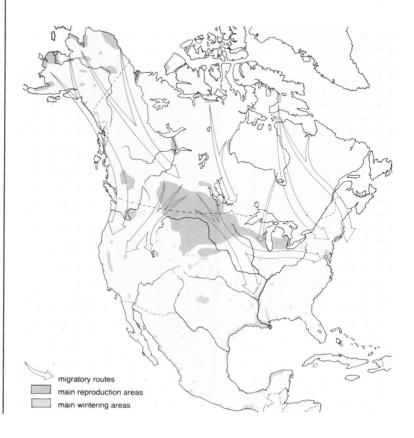

This map shows the principal routes taken by migratory birds going from their wintering areas to their nesting areas and back. How the birds, especially the water birds, manage to orient themselves along these routes still remains a mystery. Often there are no special landmarks along the routes, which may cross an open sea, plains, and so on. Only recently, scholars have suggested that birds may use the earth's magnetic field to orient themselves during their flight.

migratory routes
main reproduction areas
main wintering areas

rosy finch

Lazuli bunting
male
female

Myrtle warbler

yellow warbler
male

female

from two to twenty-two eggs in it. The female will brood the eggs and raise the young entirely alone.

A Myriad of Colors and Sounds

As spring progresses, more and more migratory birds arrive. By May, meadows and woods are full of many sounds and colors. The black-and-orange plumage of the Baltimore oriole is a common sight at this time, as well as the black-and-white plumage of the rosy finch. This bird has a bright pink triangle on its chest. As it flies, it also shows the pink coloration under its wings. The male scarlet tanager is much more peculiar. It has bright red plumage, in contrast with its black wings and ivory bill. All three species have pleasant songs. How do these showy birds avoid being seen and caught by predators? Oddly enough, their bright colors are more evident in flight. When they are perched motionless in the green treetops, they are very hard to spot, despite their bright colors.

The kaleidoscope of colors becomes richer as new birds arrive. The indigo bunting gives out its two-note call in the meadows and farmlands. Males have neon blue coloration in direct sunlight, with a darker tail and wings. The male cardinal, which lives in the same habitat, has bright red coloration, a red crest on its head, and a red or sometimes orange stocky bill. In the last few years, cardinals have become common residents, thanks to the people who leave seeds and food for them during the winter.

Undoubtedly, the most colorful and varied migratory birds of the Northeast are the warblers. These small birds are often less than 5 inches (13 cm) long. They migrate all the way to South America to winter. In spring, they are usually very colorful. Because of this, they are often called the "butterflies" of the bird world. In late April and early May, there may be as many as twenty-five different warbler species in this region. However, not all of them will nest and reproduce here. Many continue their migration further north and can be seen in flight on a nice day in May. The sight of these migrating warblers is certainly one of spring's most pleasant experiences.

Among the first warblers to appear in this region is the myrtle warbler. It has black, blue-gray, and white plumage. In contrast, the top of its head, as well as its tail and shoulders, are yellow. This species reproduces in the red spruce forests of the northern regions. The yellow warbler is another very common warbler in this region. Males have an

Despite its size, this horned owl is still too young to hunt. Perched on a log near its nest, the owl waits for its parents to bring food. Owls of this species can be quite large, growing up to 20 inches (50 cm) tall. They prey mainly on mice, voles, and other small mammals. The adults are characterized, other than by their size, by their large yellow eyes, their white throats, and two tufts of feathers, well separated on their heads. Their night call, deep and booming, is commonly heard in the woods and in the most remote areas.

intense yellow coloration, with easily visible orange stripes on their chests. The females are also yellow and are sometimes as bright as the males. Usually though, their orange stripes are less evident. The yellow warblers live in bushes, swamps, and old orchards. Here they establish a territory and then reproduce. The common yellow-throated warbler also lives in the swamps. The male of this species has a yellow underside and an olive-green back. It can be recognized by a black mask, which extends from its bill to its ears. The yellow-throated warbler is immediately spotted by its typical song, which sounds more or less like a repeated "wichity-wichity-wichity." The female has duller coloration and lacks the black mask. The yellow-throated warbler and the yellow warbler both nest close to the ground. Their nests are usually found between 3 and 6 feet (1 and 2 m) high.

Many other species dwell in the woods. The black-and-white warbler is very common and easily recognized by its black-and-white-striped plumage. Males have a black throat, while females have a white throat. This species feeds on

insects and other invertebrates which it finds on the bark of trees. These birds can climb up and down trees with ease, much like the nuthatches also common in these forests. Despite their tree habits, the black-and-white warblers nest on the ground.

Many other birds dwell in the treetops. One of them is the Blackburnianan warbler. This warbler has a white-and-black back and wings, and black-and-yellow stripes on its head. Its throat is a bright orange and looks like a burning coal when the sun's rays strike it. This bird prefers to nest in spruce trees and red spruce woods at high elevations. But it is also found in the Appalachian Mountains. Unlike the species described above, this warbler builds its nest in the treetops.

Not all forest birds have such bright colors. Some have a melodious song instead. The wood thrush is certainly one of the best singers in the forest. About 7 inches (18 cm) long, this bird has rusty hind parts and a light, almost white, breast with showy black spots. Its beautiful song is a varied succession of flute-like notes, mingled with whistles and trills. A closely-related species, the willow thrush, also has a rusty back but lacks spots on its chest. Its song is a sequence of little distinct notes which get lower and lower. Oddly enough, this sound resembles the whistle produced by blowing into a metal tube. Both species build a primitive nest. An outer mud structure is filled with grass and small roots. The nests are built on the ground or no higher than about 3 feet (1 m), and usually contain four or five blue eggs.

The birds described so far are only some of the over 150 species of birds which reproduce in the Northeast. About 60 percent of these birds are migratory, and their arrival in spring brings new life and beauty after the long winter. Songs and calls fill the woods in the first hours of the day, in the late afternoon, and in the first hours of the evening. At night, plovers and frogs take over, and the spring symphonies last twenty-four hours a day. Even in the very early spring, the call of the nocturnal birds of prey can be heard. The horned owl's deep, booming call, for example, is heard in the first months of winter. It will often be heard throughout the winter, until springtime. This large owl is 20 inches (50 cm) long, and has a wingspan of 5 feet (1.5 m). It preys upon rodents, birds, and reptiles. It lays its eggs very early in March.

MAMMALS OF THE FOREST

When spring comes, frogs and toads croak in the swamps, fish lay their eggs, and reptiles wake from their long hibernation. Thousands of birds come back from their wintering areas in the south and fill the air with new sounds and colors. Plants start to bud, growing more lush each day. At the same time, mammals living in the northeastern woods begin to appear and become active. Mammals both large and small add to the renewed life in the forest.

Forest Dwellers, Large and Small

Perhaps the most unusual mammal of the region is the opossum. This primitive animal is closely related to the Australian marsupials, although the species living in America belongs to a different family. It can be 20 inches (50 cm) long, and weigh 9 to 13 pounds (4 to 6 kg). It has a bulky body, covered by a thin, gray fur. Its head, which is a lighter color than the rest of its body, ends in a pointed snout. It has rounded ears and its hairless, scaly tail resembles a rat's. The opossum can twist its tail around tree branches, and then swing freely in the air.

In recent years, the opossum has been extending its distribution area more and more to the north. It is now widely distributed in both farmlands and woods. Each individual has a territory, varying from 15 to 40 acres ((6 to 16 hectares). However, in autumn and winter when food gets scarce, these animals may roam longer distances. In spite of the harsh winters, the opossum does not hibernate. For this reason, part of its tail or ears can be missing, due to frost damage.

The opossum is a nocturnal animal. At night, it moves clumsily on the ground, looking for food. Its wide diet includes fruit and vegetable matter, invertebrates, the young of other mammals, birds, and carrion. During the day, it hides inside hollow logs, among buildings, or in any suitable cavity. When cornered, it pretends to be ferocious by opening its mouth. If the threat gets close, it feigns death or "plays 'possum."

Young opossums are born immature, after only thirteen days of gestation. At birth, they are about 0.4 inches (1 cm) long, and weigh about 0.06 ounces (2 grams). Up to twenty young may be born but usually there are less. Like young kangaroos, the opossums immediately make their way to their mother's pouch, where they nurse for about two months. Even after they leave the pouch, the young stay with their mother several more weeks. They follow her as

Opposite: The northeastern territories are home to the red fox. This species of red fox has perhaps adapted the best to changes brought by modern farming. It hunts at night, and hides during the day. It often digs underground tunnels in which females give birth to as many as ten pups in the spring. The red fox is a predator. In this role, it helps control the rodent populations which damage crops. Unfortunately, people have hunted and killed the red fox since colonial times.

The raccoon is certainly one of the sweetest and most loved mammals in the region. If caught when it is young, the raccoon is easily tamed. Its mannerisms are similar to those of dogs, cats, and even monkeys. The raccoon is common throughout the Northeast, except in the higher elevations. Its habits vary considerably, depending on the area where it lives. It often is found close to rivers, streams, and lakes. It is a good swimmer and climber.

she looks for food, or cling to her back wrapping their tails around their mother's. Usually, the young are born in spring and summer, and opossums can have two deliveries each year. An average opossum lives up to seven years.

The raccoon is another typical dweller. It can reach 28 inches (70 cm) in length and weigh up to 33 pounds (15 kg). Its fur is mostly gray, with darker patches. Typical of the species are the white-and-black alternating rings on the tail, and the black mask around the eyes. The raccoon is widespread throughout the region. It is especially fond of wooded areas close to streams, ponds, and lakes.

The raccoon feeds mainly at night and has an omni-

vorous diet. It eats plant matter, seeds, fruits, insects, frogs, birds, and small mammals. The Latin name *raccoon* means "washer," and the animal gets its name from its strange habit of "washing" its food in water before eating it. The reason for this is not clear.

Raccoons generally make their dens inside hollow trees. But depending on the area, a hollow stump or log may also be used. Surrounding its dens, each individual has a wide territory. Some adult males claim territories up to 2 miles (3.2 km) in diameter, but most territories are about half as large. Unlike many animals of its area, raccoons are active all year round. In colder areas, they may stay in their dens during the coldest winter days, but they do not hibernate. Three or four young are born in April or May, and stay in their den until they are two months old.

Another common mammal, the skunk, is also omnivorous and easily recognized by its black-and-white-striped coat. Skunks can weigh up to 13 pounds (6 kg) and are 12 to 18 inches (30 to 45 cm) in length. Although this species is common in many areas and despite its noticeable coloring, the skunk is not often seen. Mainly, this is due to its nocturnal habits. Only its strong smell warns of its presence. The skunk's odor is produced by glands at the base of its tail. When attacked by a predator, the skunk stands on its front legs and sprays the air with a foul-smelling substance. If hit by this spray, most predators will back off, allowing the skunk to sneak away. The overpowering smell lingers on for days. This defense mechanism is extremely effective against all predators. The only exception seems to be the horned owl, which at times still preys on skunks.

The skunk's diet includes fruit, nuts, insects, eggs, small mammals, and carrion. Like the raccoon, this animal also does not hibernate but may stay in its shelter for long periods of time. Many skunks live in underground dens, but log piles, hollow logs, and even old buildings will do for a shelter. The spotted skunk, unlike other skunk species, can even climb trees. It sometimes makes its home in hollow trees.

The black bear is one of the largest mammals of the northeastern region. The largest bears can be close to 7 feet (2 m) tall, and weigh over 440 pounds (200 kg). The eastern variety of this bear is black, with a brown muzzle and quite often a white spot on the chest. Black bears live in the thickets of forests and swamps. They are omnivorous and feed on many things including berries, roots, grubs, small

The black bear, despite its name, is usually brown in color. It is one of the few large mammals still relatively common in the Northeast. It is omnivorous, eating small animals, honey, ants, bird eggs, berries, fruit, seeds, and plants. It is easy to distinguish from the grizzly bear because it is much smaller.

mammals, fish, and even carrion. They have a special liking for honey. They will often roam for miles and miles looking for food.

Bears are generally loners. Females, however, are often seen traveling with their cubs. Most cubs stay with their mother for one or two years. During this time, the female teaches her young to hunt. When hunting, bears rely a great deal on their sense of smell. The bear's eyesight and hearing are not very good, but its sense of smell is acute. This sense not only helps them find food, but it also helps detect possible danger. The female emits a typical call, a "wof-wof" sound, to alert her cubs when danger is near. Black bears are usually very shy and avoid any contact with people. When defending their young, though, they can be very aggressive.

In winter, bears enter a kind of semi-hibernation in their dens. Dens can be located at the base of an old log, in a cave, or even in a small depression under a thick layer of dead leaves. This is not true hibernation because the bear's body temperature decreases only slightly, and the bear can thus easily wake up if disturbed. If the weather is especially mild, the bear may even go into the open and feed until the temperature drops again. In January or February, one to

The white-tailed deer is extremely common, and even today it is hunted intensely all over the territory. In spite of hunting, the deer populations have risen so much in some areas that the animals have to be thinned for the sake of survival. These elegant and shy animals live mainly in woody regions from Oregon to Ontario, and south all the way to Mexico. The white-tailed deer can be distinguished from the other deer species by its large ears, rimmed with white, and the male's antlers, which point forward.

three cubs are born in the den. They are very active and feed on their mother's milk while crouching against her warm body.

The lush summer vegetation of the northeastern regions provides food for many herbivores or plant-eating animals. The largest of these is the white-tailed deer. This deer, which is the most common hoofed mammal in the eastern deciduous forests, stands 3 feet (1 m) tall or more. The average buck weighs 397 pounds (180 kg), and the average doe weighs 247 pounds (112 kg). This deer's tail, from which it gets its name, grows about 1 foot (30 cm) long and has white underparts. When the deer is startled, it runs, holding its tail straight up to reveal the white underparts. In summer, the deer takes on a reddish color, while in winter its thick, insulating coat is grayish. The white-tailed deer is common in the most densely-wooded areas. There, it moves in small herds, browsing on twigs, bushes, and leaves. During the coldest months, however, groups of twenty, thirty, or even fifty deer may form. Sometimes, during an excursion, hikers may hear loud snorting sounds. This means that there are deer nearby, and they have smelled the presence

eastern chipmunk

northern flying squirrel

southern flying squirrel

cottontail

of people. In autumn, during mating season, the males, with their large antlers, engage in fights to conquer their harem of females. The fawns are born in spring and spend their first days hidden in thickets. Thanks to their reddish brown coats with white spots, they remain well camouflaged. Soon, though, they will follow their mother and be able to easily outrun most predators. As these deer flee, they can jump nimbly, even in the thickest forests. A deer in full flight can reach speeds of 25 miles (40 km) per hour, and jump 8 feet (2.5 m) in height and 30 feet (9 m) in length. A fleeing group is an amazing sight, especially because of their incredible leaps.

Woodchucks and Other Rodents

Smaller herbivores are also abundant in the forest. This includes many species of rodents such as mice, voles, and squirrels. Particularly well known is the woodchuck, or groundhog, as it is called. This large rodent is very similar to the marmot, which is found in the western parts of the continent. But unlike the marmot, which lives in large colonies, the woodchuck is mostly a solitary creature. It lives in its own burrow, which can be 30 feet (10 m) long and 5 feet (1.5 m) deep.

The woodchuck grows to 20 inches (50 cm) long and weighs 10 pounds (4.5 kg). Its fur is yellowish brown over most of the body but grows darker around the legs. These animals feed mainly on grass and other juicy plants. They can be found along the edges of woods, browsing in meadows, in abandoned farmlands, or along stream banks.

The eastern chipmunk is another common rodent. This small animal rarely grows more than 8 inches (20 cm) long, including the tail, and does not weigh more than 4 ounces (120 grains). Its coloring goes from light brown to rust. Light-colored stripes, which are bordered by black, mark the chipmunk's face, back, and sides. This animal runs with its tail held up straight, and makes a constant twittering sound.

The chipmunk is found in all wooded areas. It is also often seen in gardens, parks, and areas with tourist facilities where food is plentiful. Chipmunks are active during the day, leaving their burrows in search of food. They feed on seeds, vegetables, fruit, insects, and occasionally also on carrion. Food is gathered in the chipmunk's cheek pouches and carried to its underground burrow. There, it is stored for the winter. Chipmunks hibernate but may awaken and come out if the weather is mild. In this case, the food supply

Marmots are large rodents, often extremely common, similar to a large stocky ground squirrel with a short tail. The woodchuck, a kind of marmot, has a reddish yellow coloring, with a lighter underside and muzzle. It lives in tunnels which it digs in the ground.

The gray squirrel is perhaps the most well-known mammal of the Northeast. It is found from the Saint Lawrence River all the way down to the southernmost regions. Its distribution stretches westward to Minnesota and Louisiana. The gray squirrel is a known tree dweller and builds large leaf nests among the tree branches or in tree holes. There, the female gives birth to a litter of young, usually twice a year.

comes in handy. Chipmunks are highly territorial and are frequently seen chasing each other along country walls, fences, or even in trees. One chipmunk's territory is usually about 180 feet (60 m) in circumference, but it can be larger in poor environments.

Many rodents, such as the southern flying squirrel, are nocturnal. This squirrel lives in deciduous forests. A very similar species, the northern flying squirrel, lives at higher elevations, mainly in the northern coniferous forests. The flying squirrel has a fold of skin along its sides, which connects its front and hind legs. When the squirrel launches

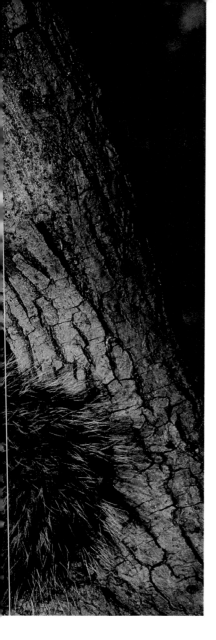

itself from the top of a high tree, it stretches its legs. The folds of skin open, forming "wings." In this way, it can glide for over 90 feet (30 m) and land effortlessly on the trunk or branch of another tree. The squirrel uses its tail to direct itself.

The northern and southern flying squirrels are the only two species of truly nocturnal squirrels. They spend the day inside tree holes, or sometimes in nests made of leaves. These squirrels are omnivorous. Like most squirrels, their diet includes fruit, nuts, insects, and eggs, but these squirrels will even eat meat. They do not store food like chipmunks and other squirrels do, but can sometimes hide a small supply in the fork of a tree or some other similar spot. They stay active all winter, but in the cold months, fifteen to twenty squirrels may share one tree hole. This habit helps them maintain a high body temperature, especially during the coldest months.

Another fairly common squirrel in the region is the gray squirrel. This squirrel is also active all year long. It is especially busy in autumn when it gathers and stores food for the winter months.

The eastern cottontail rabbit is also common in the forests and meadows of this region. This rabbit belongs to the order *Laomorpha*, which means it is not a rodent (order *Rodentia*). The cottontail is active mainly from sundown to dawn. Its days are spent in a grass-covered depression in the ground known as a "form." Many cottontails, however, will choose a better spot to spend the long winter months. Then, piles of brush or wood, or other animals' abandoned burrows may serve as home. For most of the year, the cottontail feeds on green, leafy plants such as clover and grass, but it sometimes feasts on the sprouts of crop vegetables that it finds in farmers' fields and small gardens. In winter, it resorts to chewing on bark and twigs.

Underground Mammals

A common underground dweller in the Northeast is the star-nosed mole, which takes its name from fleshy tentacles surrounding the tip of its muzzle. The mole uses these highly-sensitive feelers to locate its food. The star-nosed mole, like all other moles, has tiny and almost useless eyes, which may only enable it to distinguish light from dark. The star-nosed mole can be up to 5 inches (12 cm) long, and weighs 3 ounces (80 g). Its fur is dark brown or black, longer than other moles' and highly water repellent.

This species of mole often lives near water. It digs tunnels along riverbanks and in swamps, using its front paws, which are modified into large, shovel-like digging devices. This mole has amphibious habits, and though its hind paws are not webbed, it is a good underwater swimmer. It uses its tail as a rudder. The star-nosed mole is active all year long and can even swim below the ice. During extremely cold months, this mole digs even deeper tunnels that extend below the frost line. The star-nosed mole's actual nest is located in the deeper tunnels. There, in spring or summer, it gives birth to three to seven young.

The mole's diet consists mainly of worms and other invertebrates. As with most insect-eating animals, its metabolism is very high. In one day, it can eat a quantity of food equal to half of its weight. It usually feeds for a short time, then rests for an equal amount of time. If deprived of food, it dies within five to six hours, thus periods without food must be short.

Numerous shrews also live in the area. One of them is the short-tailed shrew, a large shrew, reaching 5 inches (13 cm) in length (tail included) and weighing up to 0.7 ounces

This drawing shows some of the main mammals living in the deciduous forests of the northeast United States and eastern Canada. Pictured (*left to right*) are: a meadow vole, a skunk, a short-tailed shrew, two deer mice, a star-nosed mole, and an opossum climbing on a branch.

(22 g). Like other insect-eaters, this animal is active day and night throughout the year. It eats almost continuously, feeding mainly on invertebrates. But it sometimes kills and eats other small mammals. For its size, the shrew is a fierce fighter and may even attack animals larger than itself, such as mice. A peculiarity of the short-tailed shrew is that its saliva is poisonous. This poison seeps into the bite wounds of the shrew's prey and kills it. This species is perhaps the only mammal in the world to have a poisonous bite.

The Gray Fox

All large predators today are found only in the northernmost areas of the region. There is evidence, however, that some species, the bobcat for example, are moving southward again. Foxes are more common and still live in great numbers in forests. The gray fox is easy to recognize. It has a grayish coat and a long, bushy tail with a black stripe on it. The fur surrounding the ears, neck, and legs can sometimes be a rusty color. This fox can be 30 inches (75 cm) long, and weigh 11 pounds (5 kg). It prefers to dwell in open woods and bushy areas in dens inside hollow logs or among the rocks.

SUMMER IN THE FORESTS

Spring activities come to an end when the trees are in full leaf. Birds are in the midst of nesting, and the loud croaking of frogs and toads is over. The forests are quieter, temperatures and humidity levels rise, and the summer heat begins. The deciduous forests of the East, with their lush underbrush and vines, look like tropical rain forests. The temperature averages 86°F (30°C). Humidity is often over 60 percent, and sometimes reaches 90 to 100 percent. Thunderstorms are very frequent, and in just a few minutes, many inches of rain can fall on a certain area. Storms cool off the air, but only momentarily. Soon the sun comes out again, the temperature rises, and due to the recent rain, the woods fill with humidity.

The Noisy Insect Community

At this time, the heat can become unbearable. This is in sharp contrast with the cold, dry winter air. During the summer, at midday, very few animals are active. Most of them are nocturnal, and do not come out at all during the day. Birds sing mostly in the early morning and late afternoon. For most of the day, they are completely silent.

The insects, on the other hand, are active day and night. Numerous species of crickets and grasshoppers produce a variety of sounds. At night, these sounds seem even louder. The fork-tailed bush katydid chirps only at night. Its sound is very familiar in the region, sounding like "katy-did, katy-didn't." This graceful insect can be 1 inch long (3 cm) and is colored green, so it is well camouflaged in the vegetation.

Without doubt, the most beautiful night show is offered by the fireflies. The firefly is about 0.6 inches (1.5 cm) long. It is black or brown, with red or yellow markings. The light organ is found at the tip of its abdomen—the last of the insect's three body parts. A series of extremely specialized chemical reactions produces flashes of bright green light. Around midsummer, many hundreds of fireflies gather in clearings, flashing their intermittent light and offering a show that has to be seen in order to be fully appreciated.

The flashes of light are used by fireflies to attract mates. Each species of firefly has its own flashing rhythm. Insects of the same species are thus able to recognize each other during courtship. Sometimes, however, a predator species imitates the flashing rhythm of its prey. The unsuspecting firefly draws near, expecting to find a possible partner and is easily caught.

Fireflies lay their eggs in moist places—often on the

Opposite: A field of yellow tickseed-sunflowers blooms at the edge of a woods in Virginia.

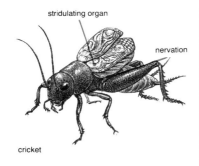

cricket — stridulating organ, nervation

cicada

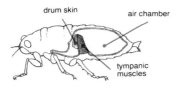

drum skin, air chamber, tympanic muscles

The concerts of crickets and cicadas are well-known sounds on summer evenings in North America. The cricket's song is produced by wing friction. A large nerve crosses the lower part of the wings, causing the vibration of a jagged organ called a "stridulating organ." This organ is located on the cricket's upper wings. The "musical organs" of the cicada, on the other hand, are located inside its body. This peculiar insect has a hollow abdomen, which acts as an air chamber. In the abdomen there are so-called tympanic muscles, which contract and tense the particular area of skin to which they are connected, called drum skin. The drum skin, tensing and relaxing, beats against the walls of the abdomen, producing the well-known summer melody of the cicada.

ground or under tree bark. The eggs hatch into flightless larvae. These larvae take one to two years to develop, feeding on snails, earthworms, and insects. The adult fireflies live for five to thirty days. During that time, they do not eat.

During the day, countless insects fly in the woods. One of the most visible of these is the large dragonfly. This flying insect has four large wings with wingspans reaching 6 inches (15 cm). Its long, slender body measures about 3 inches (8 cm), and in the sunlight, it may gleam red, green, or blue. The males of this species are strictly territorial. These shiny-bodied dragonflies are often seen flying swiftly about, like miniature biplanes, as they survey their territories.

The chirping of crickets and grasshoppers is sometimes completely outdone by the loud rasping of cicadas. These heavy-bodied insects are about 1 inch (3 cm) long and have a blackish color with green spots. The cicada's large head is accented with wide, compound eyes and short antennae. Cicadas are known for their buzzing sound. However, only the males can make this sound, which is used to attract females or call many cicadas together. The entire life cycle of most species lasts two to five years. The adult cicadas, however, live for only a few days. This time, during which they do not eat, is just enough to mate and lay eggs. These eggs are laid on twigs, in holes bored by the female. The nymphs, which hatch from the eggs, fall to the ground, burrow into it, and feed on roots for nourishment. When the nymph is fully grown, it emerges from the ground and climbs onto a tree or other nearby object. There it undergoes a metamorphosis, ridding itself of its larval sheath to become an adult.

Some species, such as the one called the "17-year locust," take between thirteen and seventeen years to become adults. The nymphs spend this entire period in the ground. Usually, these cicadas emerge all at once as if perfectly timed. In some areas, thousands of these cicadas will start rasping together in their trees, house walls, and elsewhere. Sometimes their emergence is so massive that roads and gardens are completely covered by cicadas.

A great many species of bees and wasps are also to be found in this region. One of them, the mud dauber, is very common around houses and barns. It is a large wasp, about 1 inch (3 cm) long and beautifully colored with dark blue metallic tints. This wasp builds a nest composed of parallel cylinders, resembling organ pipes, and mud which it gathers from around puddles or near streams. It carries the mud

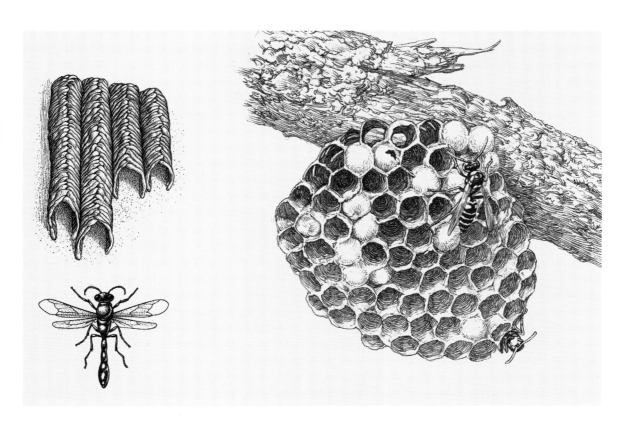

Wasps frequently build peculiar nests. Among the most unusual is certainly that which is shaped like organ pipes. This nest is built with mud by a solitary wasp. After building the nest, the wasp places an egg and a spider in each cell. This will provide food for the larva during its growth. Certain species of wasp create paper nests *(right)*. Nests like this are very common, and are frequently seen hanging from the roofs of houses. They are divided into small cells. Inside each one, the queen wasp will lay one egg. She then will take care of the larvae until they are fully grown. The queen wasp builds its nest alone, and tends the larvae until the first workers are adults.

Following pages: The sugar maple displays its gorgeous autumn colors.

to the nest in its jaws. It then lays one egg in each cylinder and adds a rich supply of spiders which it has previously paralyzed with its poison. The cylinders are then sealed with mud. When the larva hatches, it will feed on the paralyzed prey.

Another wasp species, called the "paper wasp," builds a series of paper tubes linked together. Inside each tube the wasp lays an egg, together with insects or spiders which will serve as a food supply for the larvae. This wasp makes the paper for its nest by chewing wood and mixing it with saliva. After laying its eggs and storing the food supply, the wasp seals the tubes.

The Bane of the Forests

Before ending a description of this region's insects, the so-called bane of the forests, or the deerfly, must also be mentioned. This dipteron, or two-winged fly, is similar to the horsefly. It is 0.4 inches (1 cm) long and has blotched or banded wings with iridescent tints. The deerfly can fly at high speeds. It uses this ability to chase its mammal victims, which include people.

Only the female fly, however, bites. She then sucks blood from her victims, using it to complete the ripening of her eggs. The larvae live in the water. The bite of the deerfly is very painful and can cause large skin ulcerations. Perhaps even more annoying than the bite is the constant, loud buzz of large numbers of these flies. They fly dizzyingly around one's head, briefly landing on a nose, eyelid, lip, or ear before choosing an exposed spot to bite. These insects, and others with similar habits, such as mosquitoes and black flies, are certainly annoying, but they are nonetheless an important part of the ecosystem. Their aquatic larvae are a source of food for other invertebrates, fish, and amphibians. The adults are easy prey for many species of insect-eating birds.

Autumn Colors

In September, summer heat begins to lessen, and the nights grow longer and cooler. By the end of the month, a first frost may appear. Insect numbers are also declining by this time, including that of the blood-sucking dipterians. The birds have molted, changing their bright mating patterns into duller colors. Many of them will soon move south for the winter. Frogs and toads are silent now. They are busy digging themselves into the mud as they prepare for hibernation. Many mammals are also preparing for winter. Some, such as squirrels and chipmunks, gather food supplies, while others accumulate huge quantities of fat to get them through the long winter hibernation. Vegetation takes on yellow and brown colors, and many plants go to seed and wither.

The most visible changes in the forests occur when leaves take on the kaleidoscope of autumn colors. Some species, such as oaks and maples, turn a thousand shades of red. Others turn yellow, golden, or orange-brown. These colors are already present in the leaves, but they are masked for the majority of the year by the plant's green pigment, chlorophyll. Chlorophyll is an important part of the plant's photosynthesis process. During this process, the plant uses chlorophyll to absorb solar energy, and synthesize sugars from water and carbon dioxide contained in air. As autumn begins, however, the chlorophyll in the leaves begins to decompose. As this happens, the hidden colors begin to show. At the same time, certain chemical by-products accumulate in the dying leaves. These by-products also contribute to the autumn colors of the leaves.

WOODS IN THE NORTHERN REGIONS

The deciduous forests of the northeastern regions of North America give way, at higher elevations, to conifer forests. This usually occurs at elevations above 1,300 feet (400 m) in the mountains of New York, New England, and Canada. These conifer forests lie in a great band that stretches for over 1,900 miles (3,000 km) from the Atlantic Ocean inland to the northern plains.

The Trees

Two common species of forest trees are the eastern white pine and the red pine. Both trees can grow to 98 feet (30 m) tall when the environment is favorable. Another species typical of the region is the jack pine. This tree, however, rarely reaches 80 feet (24 m) in height. This pine grows in poor, sandy soils and is usually the first species to grow back into an area after a fire. The tamarack, which is a species of larch, also grows to be 80 feet (24 m) tall. This cone-bearing tree is characterized by short, greenish blue needlelike leaves and scaly bark. Most conifers in the region are evergreen, and thus do not lose their leaves each season. The tamarack, however, is a deciduous tree. Its needles turn reddish brown in autumn, and new ones will sprout in spring. The tamarack grows on moist soils, and tamarack swamps are typical of the northern regions. Numerous species of spruce also thrive in the northern regions. Of these, the white spruce and the black spruce are the most common. The white spruce reaches 75 feet (23 m), while the smaller black spruce is generally only 40 feet (12 m) tall. The white spruce is more commonly found near streams, rivers, and lakes, while black spruces are common in wetlands.

Small woods of quaking aspens mingle with conifers in the forests. The foliage of these trees quivers in the slightest breeze, hence their common name. Another tree, often found with the aspens, is the paper birch. This tree gets its name from its bark, which peels off in long strips with a papery texture. Both species can grow to be 80 feet (24 m) tall, but they are often smaller. These plants are also among the first to colonize an area after a fire has destroyed the primary forest.

In the north, the climate is cold. Winter temperatures often fall as low as -22°F (-30°C), and around midsummer, temperatures average 68°F (20°C). Snow piles up 5 feet (1.5 m) high or more, and spring comes late, usually in May. The short summer is over by the end of August or early September. Unlike the deciduous forests of the southern

Opposite: The beaver is widely distributed throughout Canada and the United States. Because of its tree-cutting and dam-building habits, this animal has played an important role in the history of North America. In fact, millions of beavers have contributed to the formation of huge stretches of incredibly fertile soil. The beaver, which is the largest rodent of the region, can be about 3 feet (1 m) long.

A female Canadian grouse rests in the snow. The notes of this bird's call are so low that some people cannot hear them.

regions, summers here are not as hot and humid. For this reason, the vegetation is not as lush.

The Birds

Few species of amphibians and reptiles can live in these forests, but birds are well represented. A typical resident species is the gray jay, a member of the Corvidae, or crow, family. The gray jay can be up to 10 inches (25 cm) long. It is completely gray, except for a lighter area on its head and upper chest. The young, however, are much darker than the adults. Gray jays, also known as "Canada jays," have rounded wings and long tails. They are also

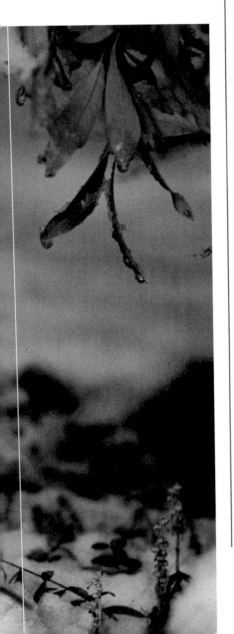

known to have a special way of flying. They flap their wings a few times, then briefly glide. These birds stay in the northern regions all winter, storing food in hideaways in trees, inside logs, and elsewhere. They pay regular visits to cultivated fields and do not fear people.

Another member of the Corvidae family, the raven, is also common in this region. This bird can be up to 26 inches (66 cm) long. The raven is very black in color, but its feathers sometimes have a purple luster to them. It has long wings, a wedge-shaped tail, and a characteristically shaggy tuft of feathers at its throat. The raven is a rather social species, often gathering in groups near sources of food. Unfortunately, these food sources also include dump sites. Its hoarse croak is a familiar sound in northern regions.

The Canada spruce grouse lives in the thick of the conifer forests. Both the male and female grouse of this species are predominantly brown with speckled sides. The male's breast, however, is sharply defined in black, and during the mating season, a bright red comb appears above each of his eyes. With this comb and a special courtship dance, the male will attract the females. Like the ruffed grouse, though, the male spruce grouse will take only one mate. Many male grouse take a harem of two to fifteen females. As it courts, the male must also mark its territory. This is done with a deep, low call. The tone of this call is so low that many people cannot hear it. Often, all that is perceived is a kind of vibration. The grouse's dull coloring serves as an excellent camouflage. When the bird is still, it blends perfectly into the background. This is especially important when there are chicks to be protected.

The Mammals

The mammals of the northern regions are no less interesting than the birds. The snowshoe hare is common all over. This animal, like all hares, is related to the rabbit. It has exceptionally long ears and strong hind legs. Because of its powerful legs, the hare will often try to outrun its enemies with great leaps. The rabbit will usually try to hide from predators. The snowshoe hare, in particular, is a brownish gray color during the summer. But in the winter, it becomes snow-white.

Snowshoe hares are well known for their population fluctuations. Drastic highs and lows occur in cycles that peak every eight to twelve years. This cycle varies in differing regions, but within the same population, the number of

The porcupine, found from the Canadian forests to Pennsylvania, is certainly an animal which does not go unnoticed. It uses the quills on its back and tail for defense. The quills, which are only loosely attached to the porcupine's skin, easily penetrate any predator's flesh.

individuals increases and decreases regularly. Many reasons have been given to explain the phenomenon of this cycle: illness, predation, climatic changes, and fires in the forests. Each one of these hypotheses provides a partial explanation, but none completely explains it.

Rodents are also well represented in the northern regions, and here some of the largest species are found. The tree porcupine is one example. This animal, which can be 20 inches (50 cm) long, and weigh 26 pounds (12 kg), is a blackish color, with long, lighter bristles. It has a huge quantity of long, sharp quills, especially on its back and tail. These quills detach easily and are used as a defense weapon. When attacked, the porcupine turns its back on its predator, threateningly waving its quills. This action produces a warning sound. If the predator persists, the porcupine will strike out with its quilled tail. In some porcupines, each quill's tip is covered with barbs. Once the barbs have pierced the predator's skin, the quills are very difficult to pull out. Instead, they work themselves deeper and deeper into the predator's flesh. For obvious reasons, most predators avoid porcupines.

A female porcupine, its quills lifted on its back, fights to protect its young from a marten. Once the quills have penetrated the predator's flesh, they tend to work in deeper and deeper, sometimes even reaching a vital organ.

Porcupines are usually solitary animals, but several may share the shelter of a stump, hollow log, or small cave for the winter. They do not, however, hibernate. Mating occurs in autumn. The females give birth, usually to a single offspring, in the spring. The young porcupines have quills at birth. These quills are soft, but they harden in a few minutes. Within a few hours, the young porcupines are already able to climb trees and can feed on solid matter, including bark.

The beaver is the largest rodent in the region. It can be 30 inches (75 cm) long and weigh 59 pounds (27 kg). The beaver has intense brown fur, but its tail is hairless and flattened. The noise created by the beaver's tail being slapped on the water is a very common sound in the regions where the beavers dwell. This sound is used as an alarm signal for the colony.

Beavers are found in streams, rivers, and freshwater lakes near wooded areas. These skilled architects build dams across the rivers and streams with inexhaustible energy. These dams can block the flow of water for several hundred feet, creating the perfect habitat for a beaver colony. Beavers use their huge front teeth, or incisors, to cut down trees and strip them of bark and branches. Some of a

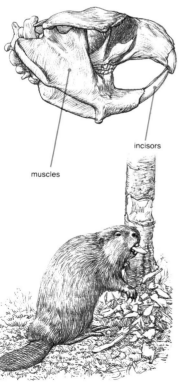

tree's branches will then be stored in deep water for winter food. The rest will be used to build or repair the dam and lodge.

Beavers are mainly nocturnal. They feed on the bark, twigs, and leaves of trees such as aspens, maples, and willows. They also eat water plants and are especially fond of their roots. A colony is usually formed by one family, and its size may vary from just a few to twelve beavers, depending on the number of young. The group consists of one adult male and female, the young from the latest litter, and those from the previous year.

Another water rodent found in the region is the muskrat. Smaller than the beaver, the muskrat is about 14 inches (36 cm) long not including its tail, which adds another 10 inches (25 cm). It weighs up to 4 pounds (1.8 kg) and has brown fur. Like the beaver, the muskrat also has a tail that is long, hairless, and flattened. Muskrats do not build dams, but instead live in burrows built in the banks. These burrows may stand up to 3 feet (1 m) higher than the water level.

The largest hoofed mammal of the northern forest region is also the largest member of the deer family. It is the moose, which can be 6 feet (2 m) tall. The male moose,

Opposite (above): The engineering work of a beaver colony is amazing. Often, the presence of these animals reshapes the landscape. Dams and pools are created, and trees are cut. The beavers will even dig channels to transport the lumber they need for their construction.

Opposite (below): This drawing shows the features of a beaver's skull, which is equipped with powerful jaw muscles and strong, continually growing incisors. Because of these features, the beaver is able to fell trees.

Above: The muskrat *(top)* is also an amphibious rodent which builds semi-flooded burrows *(bottom)*. These burrows, though, do not reach the perfection and size of the beaver's dwellings.

called a "bull," can weigh over 1,170 pounds (530 kg). The female, called a "cow," weighs about 800 pounds (360 kg). This bulky mammal has brown fur and very long legs. In autumn, the male develops massive webbed antlers which it uses in fights to obtain females for its harem. The moose has a square face and a drooping upper lip. A large fold of skin and fur, called a "bell," hangs from its throat. Moose, which live in areas of dense forests, are rarely found far from water.

The Predators

The wolf's deep howl is one of the most well-known sounds of the northern woods, and it is the symbol of nature still intact. Compared to other predators of the northern woods, the wolf is a widely-distributed species, but it is common only in the most remote areas. This animal, belonging to the Canidae family, can be up to 5 feet (1.5 m) long and weigh 120 pounds (54 kg). Wolves are mostly gray, but their coloring can vary from white (in the northern regions) to black. Wolves are social animals, living in packs of ten or more individuals. They separate only during the mating season, which occurs during the winter. After that, both parents care for the six or seven pups, which are born in April or May. By fall, the pups will be old enough to hunt, and they and the adults will then hunt as a pack.

Wolves hunt in wide territories that may be up to 62 miles (100 km) in diameter. The pack will not allow other wolves to hunt in it. To warn intruders, the leader of the pack marks the territory's boundaries with its scent. This and the pack's howling serve as a warning to other wolves to stay out of the area. Wolves prey on both birds and mammals. When they hunt as a group, they can surround and kill large mammals, such as deer or even moose.

Another typical northern predator is the lynx. It belongs to the cat family but is smaller than other wild cats such as the mountain lion or leopard. The lynx can be up to 35 inches (90 cm) long, and weigh 30 pounds (13.5 kg). Its coat is light gray or brownish gray and is speckled with a darker color. The lynx has two distinguishable characteristics: its tail and its ears. The tail is very short and has a completely black tip. Each of its pointed ears has a black tuft at the tip. The lynx is mostly solitary and nocturnal. It hunts birds, rodents, and many small animals, but it is known to feed mainly on snowshoe hares. In fact, the numbers of lynx are closely related to the hares' population cycles.

WINTER

Winter is long in the Northeast. Temperatures drop below -22°F (-30°C). In the north, at higher elevations, snow storms can drop over 6 feet (2 m) of snow, and the strong winds heap it up even higher. Spruce and aspen branches bend into graceful arches under the weight of the snow which piles on them. Some may break under such a burden, while others will spring back into position, thanks to sudden gusts of wind which free them of the snow. However, wind, too, can do damage, causing humidity to freeze on tree branches and trunks. Frost then splits the wood, with cracking noises that pierce the air like a gunshot. The surfaces of pools and lakes freeze, and blocks of ice rub against the shore making sinister noises. Below the ice, fish still swim freely, but they are not as active as in summer. Beavers venture out of their warm lodges to feed on the bark and twigs which they stored in the summer.

The situation is different on land. Many species of birds have already migrated south, and only a few will stay in the region for the winter. Marmots, chipmunks, and bears are fast asleep, hibernating inside their caves. Salamanders, snakes, and turtles have long retired to their shelters for the winter. Many of these animals will not be seen until the following spring.

Life Under the Snow

But what happens to those that remain active during the northern winters? How do they survive the intense cold and high snows? Oddly enough, snow keeps many of these animals alive. Because it is an excellent insulator, the snow provides shelter and warmth. When it is over three feet thick, it insulates the ground from the cold air. When the air temperature reaches -40°F (-40°C), the ground temperature rarely drops below 23°F (-5°C). Under the layer of snow, voles dig tunnels in search of their food. Their winter diet includes roots and other plant matter, some of which they stored in the fall. Heaps of leaves and grass under a bush are an excellent hideaway for them during the winter. Other species, such as shrews, also dig under the snow, looking for insect larvae and nymphs. In this way, these tiny creatures can remain active all winter, protected in an environment which is much less hostile than the surface.

By the end of the winter, the layer of snow is at its highest. The air can hardly circulate under this snow cover, despite the ventilation tunnels which open onto the surface. As the amount of oxygen decreases, carbon dioxide, which

Opposite: Among the rodents which stay active even during the coldest winter month is the northern flying squirrel. Despite its name, the flying squirrel does not really fly. Actually, it performs long, acrobatic glides, which allow it to move quickly from one tree to the next. When it leaps through the air, it spreads its legs open, spreading the large folds of skin attached to them. In this way, the squirrel becomes a living parachute.

is given off by the animals living beneath this layer, begins to accumulate. Carbon dioxide, you may remember, is a by-product of respiration. As living things breathe in oxygen, they give off carbon dioxide. Since carbon dioxide is heavier than air, it concentrates in the lower levels of the tunnels dug under the snow. Eventually, the level of carbon dioxide becomes so high that the animals almost suffocate. Shrews and voles are forced to go to the surface to breathe. At this point, predators finally get their chance.

The weasel and the Richardson's owl are two predators of the northeastern woods. Despite the bitter winters there, both animals manage to survive. They survey the snow-covered areas, looking for voles, shrews, small birds, and other animals which do not hibernate.

Another winter dweller of the North American woods is the ruffed grouse. This bird, which is common in the woods of the Northeast, is slightly smaller than a chicken. Males and females are quite different in appearance, but both have a rust- or gray-colored plumage. When spotted, these birds are easily recognized by their fan-shaped tails which have a black band near the feathers' tip. The bird pictured here has been caught in the middle of its mating ritual. To attract a female, the male grouse rhythmically beats its wings on a log.

Some nocturnal birds of prey, like the Richardson's owl, will catch these animals when they come to the surface. Other predators, such as the coyotes and red foxes, will crack the icy crust on the soil to capture their victims. Another, smaller predator is the weasel. This animal, which belongs to the Mustelidae family, has its own methods of hunting. The weasel is about 12 inches (30 cm) long, and usually weighs less than 6 ounces (170 gr). Because of its size, the weasel can hunt the voles right in their own tunnels. Its long, sinuous body allows it to sneak into the narrowest tunnels to reach its victims. It kills its prey with its sharp teeth, perforating the animal's skull with two strong bites. Often, the weasel kills more animals than it can eat. Some of its victims, however, will be taken back and stored in its nest. This nest, which is found in many environments, may be made in a pile of rocks, a tree stump, or an abandoned burrow. Most weasels have brown or reddish bodies with white or yellow fur on their undersides. Weasels of the colder climates turn completely white in the winter except for the tips of their tails, which stay black.

Adaptation to Life in the Snow

Some animal species seem to be at ease on the snow's surface. The snowshoe hare, for example, has large, thickly furred hind feet, which allow it to move easily on powdery snow. Thanks to its powerful hind leg muscles, this hare is an excellent jumper. With several high jumps, it can easily escape most predators. As well, large, furry hind feet prevent slippage on icy surfaces. Unfortunately, there is one predator that is equally well adapted to easy movement in deep snow. That predator is the lynx, whose diet consists mainly of snowshoe hares. Like the hare, the lynx also has stocky legs. These legs allow the lynx to run fast in the snow. So, unlike other predators, the lynx can easily catch snowshoe hares, chasing them and jumping on them, even on steep slopes.

Below the conifer trees, the snow cover is low due to the shelter of the tree's canopy. Sometimes, the ground directly below one of these trees is almost bare of snow, yet the ground all around it is covered by a deep layer. In this way, a depression is formed. Several types of animals use these depressions as shelter. Among them, there is the snowshoe hare, its predator, the lynx, and other species such as the partridge.

Not all animals have a difficult time in the deep snow

snowshoe hare

Canada lynx

The snowshoe hare's fur can vary in color from gray in the summer to snow-white in the winter. This color change camouflages the hare against the snow, thus aiding its survival during the winter season. The shape of the hare's feet also makes its movements on the snow easier. Its feet are quite large and coated with thick fur. The lynx has the same kind of foot adaptation, and snowshoe hares actually are among its most frequently-caught prey.

of the northern woods. Moose can stand cold climates quite well. They usually live near willow groves, mainly feeding on their twigs in the middle of winter. The moose's thick fur insulates it, and thus it can tolerate the most intense cold. Some nights, when the temperature drops to -40°F (-40°C), the moose remains completely still, enveloped in a slightly warm vapor created by its own body warmth. When the animal moves, it leaves this little cloud behind, while another soon forms. Moving moose thus leave a wake of vapor behind them.

During the winter in the more southerly regions, the temperature will quite often rise and thaw the surface snow. Soon after, however, it refreezes, forming an icy crust. Moose and deer, which are heavy animals, often crack the surface ice, sink in the underlying snow and cut their legs deeply on sharp edges of ice. Bloody spots are not a rare occurrence along deer tracks.

Many other species have developed ways to endure the intense cold of this region. Squirrels, for example, dig in the snow to retrieve seeds and pine cones which they hid there in the autumn. At night, they seek shelter in nests made of leaves inside tree holes. Birds such as gray jays and titmice also store their food and use it when normal supplies are scarce. Titmice and other small birds spend the night gathered in groups. Perched in the thick of conifers or other trees, the birds are protected by their highly-insulating plumage. When it sleeps, each bird hides its head under a wing, looking like a feather ball with only a tail sticking out. In winter, the jays also are covered by a particularly thick plumage which insulates them from the cold very well.

Some other species, like the red crossbill, can even reproduce in northern region winters. This species moves a great deal and nests when it finds conifer trees with seeds on them. The male has a dull, red-colored body. Its tail and wings are a darker red or black color. The female is olive green, with a yellowish underside. The bill of this bird is very distinctive. The tips of the upper and lower beak are not aligned as are most birds' bills. Instead, they cross over each other. This particular adaptation proves useful for separating the scales on pine cones and taking out the seeds.

Pine cones are a very nutritious source of food for the crossbills. They play an important role in its reproduction process. When the birds find large quantities of pine cones, they begin to reproduce immediately, regardless of the sea-

The American wolf, belonging to the same species as the European wolf, is found today in only the most unspoiled regions. Even in these places, it is present in rather small numbers. Unlike many other predator mammals, this animal is unable to withstand competition with people. Wolves live in family groups, which continually range about in their vast territories. Sometimes, during the winter, they may gather in more numerous packs. Within the northeastern regions, the color of their coats can vary considerably, from white and gray, with a creamy belly, to red and black.

son or temperature. The nests the crossbills build are large in comparison to their size. Crossbills are usually about 5 inches (13 cm) long, while their nests are up to 9 inches (23 cm) in diameter and very deep. These nests must also be well insulated in order to keep the eggs warm. When females brood, their bodies fit perfectly into the inner shape of the nest, thus completely insulating the eggs from cold air. As a result of this adaptation, there is evidence of eggs hatching in January when air temperatures were -31°F (-35°

Some migrating birds, like the junco pictured here, have become residents in the northernmost regions. During winter, these small birds poke through dead leaves on the ground or on tree branches with their bills, looking for seeds and small insects.

C). Obviously, the female can never leave the nest because its eggs or nestlings would freeze in a few minutes. In such a situation, the male crossbill plays a very important role. It must feed the female in the nest. The local abundance of pine and spruce seeds makes this job easier than it might otherwise be, thus making reproduction possible even under extreme weather conditions.

Winter in the Deciduous Forest

As these examples have shown, winter in the northern regions, as cold as it can be, is a perfect season for some species. Even more to the south, in the area of the deciduous forests, there is heavy snowfall. Here, though, the snow does not remain as long as it does in the northern woods. In winter, many more species stay in the deciduous forests. Many of these animals take advantage of the thick vegetation that grew the previous summer. Although dry and withered, this vegetation makes good shelters, even when the snow is piled high.

When snow falls, it accumulates on top of bushes and dead leaves, leaving free spaces below. There, many birds,

This drawing shows some of the birds which live in the woods of the Northeast, even in winter. They are (*left to right*): the white-winged crossbill, the titmouse, the white-throated sparrow, the gray jay, and a pair of red crossbills.

such as the white-throated sparrow and the slate-colored junco, can find seeds, gathering them directly from the ground. This situation changes only when some very heavy snowstorms hit the region. If the temperature is very low, powdery snow will result. This type of snow seeps through the leaves, especially if the storm is accompanied by strong wind. Such winds sweep this light snow around in all directions. When this happens, birds have a hard time finding food. They are forced to dig through the snow to find it. As the snow layer gets thicker, the energy required for digging increases, until the effort is greater than the reward.

It is at this point that many birds move, sometimes leaving the area simultaneously. Many of these birds have fat reserves that, if necessary, could be used for migration toward more hospitable southern regions. However, it is not certain that all of the populations move south. This is due, in part, to an increase in food sources available to these birds. In the last few years, many people have placed bird feeders in their gardens. They keep these feeders well stocked with seeds all of the time. Because of this trend, birds which used to move southward now simply move to the nearest bird feeder.

THE ATLANTIC COAST

Long stretches of northern coast are rocky, and cliffs abound along the shores. The weather is cold all year long. Summer temperatures are rarely higher than 86°F (30°C), and in the winter, it always falls below freezing. However, it rarely grows as cold here as it does inland. Because of the ocean's influence, the sharp temperature drops typical of inland regions are not common here.

Atlantic Whales

Many species of whales, which are some of the largest mammals in the world, are found in the Atlantic Ocean. One of them, the humpback whale, was once abundant in these waters. About fifty years ago, however, it was hunted nearly to extinction. As recently as 1970, this whale was included on the list of endangered species. But today, its numbers are growing again.

The whale populations of the northern Atlantic Ocean spend the winter in the Caribbean Sea. There they mate and reproduce. These tropical waters, though, are not rich in plankton, which is a primary part of their diet. So, while in the tropics, adult whales must live off their blubber reserves. The whales spend the summer building and storing these reserves. In spring, humpback whales migrate all the way to the Grand Banks, where they feed on herring, squid, and small crustaceans known as "krill." They also eat large quantities of young fish known as capelin.

Scientists divide whales into two major groups: baleen whales, which do not have teeth, and toothed whales, which do have teeth. The humpback whale, which belongs to the baleen group, has a huge, toothless mouth. Instead of teeth, it has long rows of thin plates called baleen. About three hundred of these plates hang from each side of the whale's upper jaw. These plates, which are made of fingernail-like material, have fringed edges. As water flows through the baleen, the fringed edges filter out masses of plankton. The filtered food is then swallowed.

Humpback whales can be up to 50 feet (15 m) long. They are easily recognized by their narrow fins, which are about one-third of their total length. They can dive for many minutes, and when they surface, the whales exhale. This action produces the well-known cloud of water vapor known as the "blow" or "spout." Humpbacks swim in small groups, communicating with each other through an amazingly rich variety of sounds. These sounds include grunts, moans, and sharp-pitched whistles. Humpback whales are

Opposite: Immense colonies of seabirds are typical of the rocky coasts and coastal islands of the North Atlantic. These colonies may be formed by several species, or by a single one, as is the case in this picture. This colony, made of thousands of gannets, can cover the entire coast.

Seeing a group of humpback whales swimming on the ocean's surface is a moving experience. They emerge from the water and hit it with their powerful fins, causing splashes of water to fall several feet away. This show is rare, but it can be seen from various towns along the coasts. In many places, boat excursions are organized to observe the large ocean fauna. Humpback whales, like all other species belonging to the baleen group, feed by filtering large quantities of water through their baleens. This species has a very complex "song," with some of the highest-pitched and lowest-pitched notes in nature.

famous for their sounds, which are often called their "songs." These songs, which may be very loud and varied, certainly have some kind of social meaning, but scientists do not yet know what the songs convey.

Other species of whales also roam along the Atlantic coast. Among them is the fin whale, which may grow up to 69 feet (21 m) long. These whales sometimes come close to the coast to feed. There also is the blue whale, which is often seen along the coast. The blue whale is the world's largest known living animal. It measures over 98 feet (30 m), and weighs over 300,000 pounds (136,000 kg). It usually has a bluish gray back, and white or yellow coloring on its belly. It

feeds mainly on krill, which it filters through its massive baleens.

Recently, a small population of right whales was also discovered off the Bay of Fundy in the north Atlantic Ocean. The right whale was once common all over this ocean. But since 1920, it also was heavily hunted. Like the humpback whale, this whale was on the verge of extinction. Today, the whale is protected. After many years of protection, the right whale is still rare. For this reason, the recent discovery of about sixty specimens was a pleasant surprise. This whale, which can grow up to 69 feet (21 m) long, is a slow swimmer and a very docile animal. Both qualities make the whale easy to catch. It feeds on krill close to the coast, and is easily recognized by the typical fleshy bumps on its head. It is thought to migrate toward tropical waters in winter, but some individuals remain in the Bay of Fundy throughout the year.

Colonies of Seabirds

The northern coast is well known for its hazy crags, inhabited by huge numbers of nesting seabirds. The solan goose, or gannet, is a large seabird. It is often 31 inches (79 cm) long, with a wingspan of 71 inches (180 cm). The gannet has a white plumage, with black wing tips and a yellow head. It feeds on fish and nose-dives from great heights into the ocean to catch its prey. Usually the gannet nests in densely-populated colonies along coastal islands. A colony can be seen from a distance, with the snow-white birds regularly spaced about 3 feet (1 m) from each other. They nest all over the island. Their white droppings make the surrounding rocks appear pure white.

The most rugged coasts are often colonized by murres and razorbills. Often the colonies are overcrowded, and the birds must fight constantly to keep their spots. Both species are members of the auk family and have black-and-white coloration. The razorbill, however, can be recognized by its black bill, which has a white band halfway between its base and tip. Both the murre and the razorbill nest in densely-populated colonies on rocks, where the females lay a single egg directly on the bare rock. The eggs of these birds are sharply pointed at one end, like a top. This shape makes them roll in tight circles, so they are less likely to fall from the rocks.

These sheer cliffs are also home to the kittiwake, a small sea gull with white-and-gray plumage, black wing

tips, and a grayish bill. This bird builds its nest with seaweed on tiny rock projections. Often, the cliffs become mixed colonies of auks, gulls, and gannets. Such cliffs can become very noisy places, with these birds' calls being heard several miles away.

The lobster is a well-known dweller of shallow waters along the coast. It is thought to be able to grow as long as 3 feet (1 m) and weigh 44 pounds (20 kg). But most of the time, lobsters are much smaller. Most are a bluish black color and are armed with two huge pincers. The lobster uses its pincers to fight fiercely when caught. This animal is strictly nocturnal. It comes out of its hideaways, usually located among the rocks or in underwater caves, only at night to hunt other invertebrates. It also feeds on dead fish, and these are often placed in lobster pots as bait.

Sandy Beaches and Brackish Swamps

Along the central Atlantic coast, there are long stretches of sandy beaches, behind which large brackish swamps are formed. These environments are extremely rich in plants and animals, and they attract huge numbers of migrating birds. Flocks of shorebirds fly along the coast. Often they are headed north on their way to nesting areas in the Arctic. At other times of the year, they are going south looking for tropical wintering areas. Gulls feed in shallow waters on anything which is edible.

The skimmer is a bird distantly related to sea gulls, and its feeding pattern is probably unique. It is a large bird, with a wingspan of 48 inches (122 cm), a black back, a white belly, and a red bill. The skimmer's bill looks like a pair of scissors. The lower part of the bill is larger than the upper part. When it hunts, the bird flies close to the ocean surface, with the lower part of its bill in the water. When it touches prey, usually a small fish, it bends its head downward, snaps its bill closed, and seizes its victim.

Other birds common to the coastal swamps are the wading birds. Several species of herons, from the great blue heron to the small American bittern, also are common in all wetlands. Along the drier coasts lives the cattle egret, a species which has recently moved into the region from the south. This species formerly lived only in Asia, Africa, and Europe.

The glossy ibis also lives in swamps. It has a uniform, bronze-brown coloration. It sticks its long bill, curved downward, into the silt looking for invertebrates.

Opposite (right): The American bittern has the peculiar habit of keeping its bill in line with its neck, both while it rests and while it looks for food. This habit helps young bitterns, like the one pictured here, camouflage themselves in the lake vegetation. If the bird stays still for a long time, even a predator will not notice it.

Other Peculiar Animals

Shallow bays, lagoons, and channels are the ideal habitat for a very peculiar invertebrate, the horseshoe crab. This animal has a shell shaped like a horseshoe, followed by a central, spiky part, and a long tail. It can be 23 inches (60 cm) long. It has compound eyes set directly in its olive-colored shell. Despite the name, these animals are not actually crabs. In fact, they are more closely related to scorpions and spiders than to crustaceans. They belong to the genus Limulus, which is part of a larger group of animals that date back to the Triassic period 245 million years ago. Horseshoe crabs have six pairs of legs. The front pair of legs has pincers that are used to capture and eat prey. The other five pairs are used for walking. The horseshoe crab feeds at night, preying on mollusks, worms, and other animals.

The scissor-billed skimmers are very peculiar seabirds. They often gather in huge groups after fishing, landing on the shore to rest. A few individuals, however, can still be seen darting low over the ocean surface. Keeping the lower part of their bills in the water, the skimmers fish in their characteristic way.

In late spring, the horseshoe crabs move into shallow waters to mate. The female lays between two hundred and one thousand greenish eggs in the tidal zone. There she has dug several small holes in the sand. The male then fertilizes them. Many birds prey on the horseshoe crab eggs, especially gulls and flocks of migrating shore birds, such as the turnstone. Many invertebrates also feast on them. The young horseshoe crabs are sandy colored and lack a tail. They stay in shallow waters until after their first shedding. They then acquire a tail and begin to look like adults. Only then do they start moving toward deeper waters, increasing their size, and becoming gradually darker at the same time.

GUIDE TO AREAS OF NATURAL INTEREST

Wildlife is still abundant in many regions of North America today. This is true even though the area has experienced an increase in urbanization and pollution in the last few years. Today, though, untouched environments are found only in the northernmost regions. In an effort to save what remains of the original natural environment, numerous parks, reserves, and other protected areas have been created. Many shores and coastal swamps also are being protected.

Like the United States, Canada also has an extensive national park system. In addition to this, each province in Canada also has its own park system. In eastern Canada, there are several provincial parks which cover the vast inland forests, up to the extreme northeastern coast. Many of these are located in isolated regions, with very few tourist facilities. Most of them are easily reached by car, but along the road there are very few towns or motels. Each park offers campgrounds and a number of trails for adventure lovers. Most of the visitors come in summer and autumn because winter is very cold. The temperature regularly drops below -22°F (-30°C). Even in winter, though, these parks deserve a visit.

The eastern regions of the United States have few national parks compared to the western regions. There are, however, areas of natural interest, protected as national forests or natural reserves, which are open to the public. Most of the eastern territory is privately owned and is usually off limits to the general public. This also is true of many beaches, although in most of the states, part of the coast is open to the public. State parks are numerous, but since many of them are heavily used by weekend tourists, they are not ideal for the nature lover.

The following pages briefly describe the most interesting areas. Many of the species dealt with in this book can be observed in these areas. Most of these sites can be reached by car, and motels or other tourist facilities are usually nearby. Campgrounds also are found in the region, but they are not as numerous as in the West or in Canada.

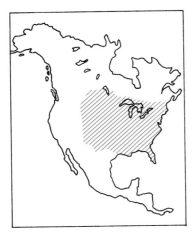

Below: The drawing outlines the area described in this book: the northeastern territories of the United States and Canada.

Opposite: An excursion into the forests of northeast America shows that some of the forests remain untouched by civilization. These areas teem with plant and animal life. Such environments offer pleasant surprises, like seeing a small brown bear bravely climb a tree as its mother watches from the ground.

CANADA

Ontario: Quetico (1)

This provincial park covers about 2,300 sq. miles (6,000 sq. km) of lakes and forests at the border between Ontario (Canada) and Minnesota. Found here are the typical habitats and animals of the northern woods. There also are numerous lakes, rich in fish life. Probably the best way to

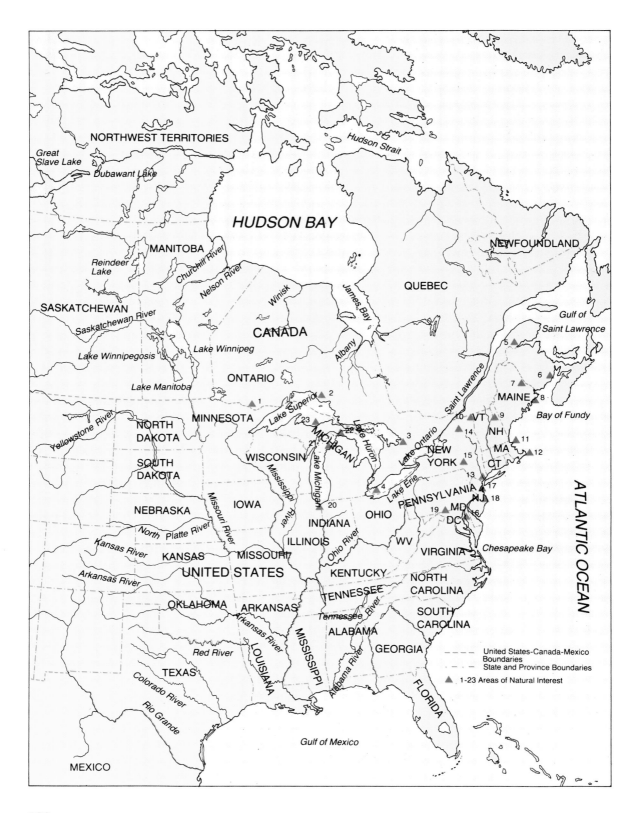

Opposite: This map shows the distribution of the areas of natural interest in the American Northeast.

**Ontario:
Lake Superior (2)**

Ontario: Algonquin (3)

Ontario: Point Pelee (4)

**Quebec:
Forillon National Park
and Gaspe Peninsula (5)**

travel inside the park is by canoe. In winter, the surfaces of lakes and rivers freeze completely. Both conifer forests and deciduous forests are found in the park. It is possible to run across moose and deer, and wolves are also common.

This very wild area is best reached by plane from nearby towns. There are no roads inside the park. In some areas, a canoe is a necessity.

This provincial park is comprised of about 155 sq. miles (400 sq. km) of forest and lakeshore. This area is rich in typical northern forests, and has breathtaking views of the largest lake in the world. Excellent excursions can be organized along the lakeshore and into the forests.

This provincial park is located in a wide area covering over 2,700 sq. miles (7,000 sq. km) of woods and lakes. This area is excellent for observing moose, beaver, grouse, and, in summer, numerous species of birds. Gray jays are common in picnic areas. The park includes numerous lakes, some of which are quite large. Fishing is generally very good, although some areas have been affected by acid rain.

This national park includes the peculiar, triangle-shaped Pelee Peninsula, which projects into Lake Erie for about 7 miles (12 km). The park territory is covered by low forests and wetlands, and is located on the northern shore of Lake Ontario. The long peninsula attracts large quantities of migrating birds. Many birds are seen flying across the lake in spring. Others pass through the area on their way south in autumn. As a result, huge quantities of migratory birds can be found here many times throughout the year. Finches, thrushes, jays, and many other species all fly toward the point. This also is a very good spot to watch ducks, both in the marshes and on the open lake waters.

The point of the peninsula can be reached from Windsor, Ontario, or from Detroit, Michigan. Two other parks along the northern shore of Lake Erie also are worth a visit. These are Rondeau Provincial Park and Long Point Provincial Park. Both parks are excellent sites for migratory bird-watching, in a beautiful lake setting.

The small Forillon National Park is 155 sq. miles (403 sq. km) wide. It is located in eastern Quebec, close to the mouth of the Saint Lawrence River. The climate here can be harshly cold, due to humid and cold winds blowing from

A tree swallow rests, well sheltered in its nest. Birds are particularly numerous in northeast America. Many different species dwell in the wetlands, and the deciduous and coniferous forests.

New Brunswick:
Fundy National Park (6)

the ocean. Local people say that there are only two seasons in the region: July and winter. The park includes long, rocky coasts, often shrouded in fog. Large colonies of seabirds nest on the crags and cliffs. On Bonaventure Island, a small provincial park to the south, it is possible to see murres, razorbills, and kittiwakes. Boat excursions to visit the most interesting bird colonies are available. Information about the trips can be obtained on the spot.

This small national park in New Brunswick is about 58 sq. miles (150 sq. km) wide. It is the ideal spot for watching tremendous tides, among the highest in the world. The peculiar rock composition of the coast is also interesting. In spring and autumn, large numbers of migratory shorebirds and ducks can be seen in the area.

UNITED STATES

Maine: Baxter (7)

This state park includes over 231 sq. miles (600 sq. km) of lakes and forests in northern Maine. Within the park is Mount Katahdin, which stands 4,921 feet (1,500 m) high, and the ancient volcanic peaks. The park is an excellent example of northern forests with all of the associated wildlife.

The northeastern corner of the protected area reaches the Hallagash Wilderness Waterway, which stretches north for over 62 miles (100 km). The lakes and rivers of these northern regions can be explored only by canoe. The only access to the area is by a private road west of Presque Isle in northern Maine. The road is open to the public with permission.

Maine:
Acadia National Park (8)

This national park is comprised of a few coastal areas and small islands near the mouth of the Bay of Fundy. Its overall surface, including all the islands, is only 115 sq. miles (300 sq. km). The natural beauty of this area is incredible, and pine and red spruce forests come down all the way to the rocky coast. Wildlife is very abundant, especially sea animals, which can be observed at low tide in the tide pools along the cliffs.

New Hampshire:
White Mountains (9)

This mountain chain, in the state of New Hampshire, is 123 miles (199 km) long. Its highest point, Mount Washington, is over 3,937 feet (1,200 m) high. Deciduous forests are found in the valleys, while the mountain slopes are covered by coniferous woods. At higher elevations, the forests disappear.

**Vermont:
Champlain (10)**

This vast lake, though smaller than the Great Lakes, touches the states of New York and Vermont. It is 40 miles (65 km) long and is fringed by conifer and mixed forests. Besides the beautiful scenic landscape, this lake also is a good site for migratory bird-watching. In summer, there are many animals and plants typical of the northeastern regions.

To the east and south of the lake are the Green Mountains of Vermont. These mountains, similar to the Adirondack and Catskill mountains, are especially well known for their gorgeous fall colors, as well as for their natural history.

**Massachusetts:
Plum Island (11)**

This state park includes a stretch of sandy dunes, brackish swamps, fresh waters, coastal woods, and open ocean beaches. It is one of the most famous areas for bird-watching in North America, especially during the migration period. Great numbers of birds can then be seen, especially shorebirds. Trails cross the swamps which stretch among dunes and flat ground, allowing visitors to closely watch birds. Many coastal towns, such as Newburyport, Gloucester, and Portland, offer special boat trips for whale-watching. The right whale and the humpback whale usually can be seen during these excursions, as well as many seabirds which are difficult to watch from the coast.

**Massachusetts:
Cape Cod (12)**

The Cape Cod National Seashore is a large territory, extending from inland Massachusetts all the way to the Atlantic Ocean. It covers over 43 miles (70 km) of coast and includes sandy beaches, brackish swamps, dunes, and coastal forests. Cape Cod can be visited any time of the year, although in summer the area may be overcrowded with tourists and short on accommodations. In winter, the weather is often cold and humid. But even then, it is possible to go for very pleasant walks along deserted beaches. Spring and summer, of course, are good periods for observing the coastal vegetation. Bird-watching is best during the migration periods. Even in winter, however, bird-watching can be spectacular. Many seabirds can be seen flying over the open sea, feeding in shallow waters or wandering on the coast.

**New York:
Tivoli (13)**

This small area of deciduous forests and freshwater marshes on the Hudson River is particularly well known for its freshwater "tidal" swamps. The ocean tides affect the river's water level for over 62 miles (100 km) upstream. Salt water, though, does not extend all the way to Tivoli. As a result, a very particular freshwater tidal environment has

In the autumn, the aspens display spectacular colors. This stand of aspens is found in the Adirondack Mountains of New York State.

been created. In summer, the swamp vegetation is very interesting, and the birds offer a beautiful show during migration. Beavers and muskrats also are found in the area.

New York: Adirondack (14)

This forest reserve includes a wide area of over 3,860 sq. miles (10,000 sq. km) in the northern part of New York state. It is a splendid region, rich in forests and valleys. Here some of the highest mountains of the Northeast are found. Numerous trails cross the park and reach the most remote spots. Due to the densely-populated areas close to the park, the campgrounds and the many tourist towns can become overcrowded. The adventure lover, though, can still find silence and peace in valleys and mountains away from the beaten track. The forests are mostly still intact and in a primary stage, but many lakes have been affected by acid rain. Many species of birds, mammals, and insects are common, and the vegetation is highly varied.

Summers in the reserve can be hot and humid. But sudden temperature drops at higher elevations are to be expected. Winters are cold, with a great deal of snowfall. Temperatures sometimes drop extremely low, with strong winds making the situation even worse. The Adirondack Mountains, though, can be pleasant even in winter. Cross-country skiing can be a great way to get around the area, and it is also a good way to observe winter in the Northeast.

The park offers many facilities for both the camper and the less daring visitor. Many hotels are open year-round. Finding a room in winter, though, may be difficult, especially in the middle of the winter sports season.

New York: Catskill (15)

This park stretches over 1,080 sq. miles (2,800 sq. km) through a woody region typical of the Northeast. It includes plains, valleys, mountains, and peaks up to 3,937 feet (1,200 m) high. Beautiful deciduous forests cover the valleys, especially along the numerous streams and rivers. As the elevation increases, the forests become mixed, and the conifers become dominant on the mountains.

There are a great many species of animals and plants to be found in this park. Some of them are typical of the northeastern plains, and some are typical of the northern woods.

Maryland: Assateague (16)

Assateague is one of the islands along the central Atlantic coast. It has dunes, coastal forests, swamps, and open ocean beaches. The island is about 31 miles (50 km) long and includes the Assateague National Seashore and the Chincoteague National Wildlife Refuge.

New Jersey: Brigantine (17)

This is one of the numerous protected areas along the coast of New Jersey. It is a National Wildlife Refuge, stretching over 11 sq. miles (30 sq. km) of brackish swamps, lagoons, and open ocean beaches. A comfortable country road winds through the entire area in a loop, offering spectacular natural sights. Large flocks of geese and ducks can be seen in winter, while in summer there are large numbers of herons, little egrets, and rails. In spring, crowds of horseshoe crabs lay their eggs in shallow waters.

New Jersey: Cape May (18)

This state park is located in the southernmost part of New Jersey, in a most beautiful sea vacation area. The buildings are old Victorian style, and the region is rich in coastal woods, swamps, and beaches. In spring and autumn, bird migration is undoubtedly more spectacular here than in any other part of North America. The migration of shorebirds and other birds makes a spectacular show. Farther inland, there are lakes and paths leading to the swamps. Here, the marshy vegetation and typical lake birds can easily be observed.

Virginia: Shenandoah National Park (19)

This national park which covers over 62 miles (100 km) of mountain area covers part of the Appalachian Mountains in Virginia. These mountains are not as high as those farther to the north, but the beauty of their forests, clearings, valleys, and crests is breathtaking. This is the perfect place to observe the southern deciduous forests.

Indiana: Indiana Dunes (20)

The Indiana Dunes National Lakeshore includes about 9 miles (15 km) of Lake Michigan shoreline. This stretch of shoreline has been maintained in a natural state. Coastal dunes predominate, offering a rich sample of the flowers and plants of the Great Lakes region. At the proper time, migratory birds can be seen flying over the area. The most interesting features of the park are its lake panoramas. It also features numerous endangered lake plants.

Michigan: Sleeping Bear Dunes (21)

Sleeping Bear Dunes National Lakeshore covers 31 miles (50 km) of forests, dunes, and Lake Michigan shoreline, as well as the two South Manitou Islands. These islands can be reached by ferryboat from Leland, Michigan. This area offers habitats similar to those that must have

Preceding pages: The waters of a pond, teeming with vegetation, shimmer in the sunlight.

Michigan: Wilderness (22)

Michigan: Pictured Rocks (23)

surrounded the lake years ago. The dunes, forests, and wetlands are worth a visit. The main features of the park are its vegetation, the migratory birds, and the beautiful lake views.

This is a small state park near Waugoshance Point, along the northern shores of Lake Michigan. It is one of the few stretches of relatively untouched forest in the entire system of the Great Lakes. Here visitors will find many species of native plants, as well as many species of birds, mammals, and invertebrates. As always in this region, the wide open views on the lake are astonishingly beautiful.

Pictured Rocks National Lakeshore covers over 31 miles (50 km) of Lake Superior's coast. The area includes forests with plants, birds, and mammals typical of the region. The park adjoins Hiawatha National Forest, which crosses the northern peninsula of Michigan, between Lake Superior and Lake Michigan. The entire region is interesting because it is surrounded by Lake Superior, Lake Huron, and Lake Michigan.

GLOSSARY

acid rain rain with a high concentration of acids produced by sulphur dioxide and other chemicals during the burning of coal, petroleum, or natural gas.

algae primitive organisms which resemble plants but do not have true roots, stems, or leaves.

amphibian any of a class of vertebrates that usually begin life in the water as tadpoles with gills, and later develop lungs.

bacteria one-celled microorganisms which have no chlorophyll, multiply by simple division, and can be seen only with a microscope.

carbon monoxide a colorless, odorless, highly poisonous gas.

carrion the decaying flesh of a dead body, regarded as food for scavenging animals.

chlorophyll a green pigment found in certain organisms that is used in the process of photosynthesis.

conifer cone-bearing trees and shrubs, most of which are evergreens.

continent one of the principal land masses of the earth. Africa, Antarctica, Asia, Europe, North America, South America, and Australia are regarded as continents.

crest a tuft, ridge, or similar growth on the head of a bird or other animal.

crustaceans marine invertebrates, characterized by a segmented body, hard outer skeleton or shell, and paired, jointed limbs.

deciduous forests forests having trees that shed their leaves at a specific season or stage of growth.

deforestation the clearing of forests or trees. This mass removal of forests was once done for agricultural and industrial purposes.

diatoms a group of single-celled plants which helps to form phytoplankton.

drumlin a long ridge or oval-shaped hill formed by glacial drift.

drupe any fruit with a soft, fleshy part covered by a skinlike outer layer, all surrounding a seed in the center.

ecosystem the relationship formed by the biological environment (which includes all living things in an area) and its physical environment.

environment the circumstances or conditions of a plant or animal's surroundings.

erosion natural processes, such as weathering, abrasion, and corrosion, by which material is removed from the earth's surface.

flora the plants of a particular region or time.

glaciers gigantic moving sheets of ice that covered great areas of the earth.

habitat the area or type of environment in which a person or other organism normally lives or occurs.

herbivore an animal that eats plants.

hibernation spending the winter in a dormant state.

invertebrates lacking a backbone or spinal column.

lithosphere the solid, rocky part of the earth.

metamorphosis a change in form, shape, or function as a result of development; the physical transformation of various animals from the early stages of life.

mimicry a condition in which an organism closely resembles, or mimics, its surroundings or another animal or plant.

mollusk an invertebrate animal characterized by a soft, usually unsegmented body, often enclosed in a shell, and having gills and a foot.

moraine a mass of rocks, gravel, sand, and clay carried and deposited by a glacier.

nocturnal active at night.

omnivorous animals that eat both plants and other animals.

parasite an organism that grows, feeds, and is sheltered on or in a different organism while contributing nothing to the survival of its host.

parr a young salmon during the first two years of its life when it lives in fresh water.

peninsula a land area almost entirely surrounded by water and connected with the mainland by a narrow strip of earth called an isthmus.

pheromone a chemical substance secreted by certain animals which conveys information to and produces specific responses in other individuals of the same species.

phytoplankton small, floating aquatic plants.

photosynthesis the process by which chlorophyll-containing cells in green plants convert sunlight to chemical energy and change inorganic compounds into organic compounds.

polymorphism the condition in which a species has two or more very different structural forms.

predator an animal that lives by preying on others.

redd the spawning area or nest of trout or salmon.

reptile a cold-blooded vertebrate having lungs, a bony skeleton, and a body covered with scales or horny plates.

rodent any of a very large order of gnawing mammals, characterized by constantly growing teeth adapted for chewing or nibbling.

sedimentary rock rock formed from sediment or from transported fragments deposited in water.

silt sedimentary material consisting of fine mineral particles close in size to sand and clay.

spathe a large, leaflike part of a plant which encloses a flower cluster.

spawn to deposit eggs. Different marine species have their own spawning periods and habits.

tadpoles the larva of certain amphibians, such as frogs and toads, having gills and a tail and living in water.

tectonic plate one of several portions of the earth's crust which has resulted from geological shifting. The earth's plates have been moving continually for millions of years, causing new surface features and geological shapes.

thicket a thick growth of shrubs, underbrush, or small trees.

tornado a violent, whirling column of air extending downward from a large cloud.

toxic poisonous.

tributary a small stream or river which eventually flows into a large body of water.

verruca a wartlike elevation, as on a toad's back.

vertebrates animals having a backbone or spinal column.

vortex a whirl or powerful air movement; a whirlwind.

weathering the physical and chemical effects of the forces of weather on rock surfaces.

zooplankton floating, often microscopic sea animals.

INDEX

Acid rain, 20-21
Adirondack Mountains, 36, 114-115
Alewife, 30
Algae, 23, 24, 25
American Falls, 13
Amphibians, 40-46
Appalachian Mountains, 9
Areas of natural interest
   Canada, 107, 108, 112
   United States, 112-113, 116-120
Arrowhead, 33
Aspens, 83, 114-115
Atlantic coast, 10-16, 99-106
Autumn colors, 81

Baleen whales, 99
Baltimore oriole, 56, 61
Basswood, 39
Beaver, 82, 83, 87-88, 89
Beech, 38
Bees, 78
Birds
   forest, 83-85
   migration routes, 60
   migratory, 57-58, 96
   nonmigratory, 58
   seabirds, 101-103
   shorebirds, 103
   wading, 103
Bittern, 102
Black-and-white warbler, 62-63
Black bear, 67-69
Black duck, 33
Black flies, 55
Black spruce, 83
Bluebird, 57, 58
Bluegill, 50-51
Blue heron, 102, 103
Blue jay, 58-60
Blue whale, 100-101
Blue-winged teal, 33
Brackish swamps, 103
Bronzed grackle, 57
Bullfrog, 43, 44
Burbot, 25, 26
Bur reed, 31
Butterflies, 52, 53

Canada grouse, 83
Canada jay, 84
Canada spruce grouse, 85
Canadian areas of natural interest, 107, 108, 112
Canadian Shield, 11
Canidae (dog) family, 89
Cape Cod, 16, 18, 19
Cattail, 31
Cattle egret, 102, 103
Chesapeake Bay, 17
Chestnut, 38
Chipmunk, 70, 72, 81
Chlorophyll, 81
Chub, 25, 28
Cicada, 78

Cisco, 25
Coastal swamp, 17
Coniferous forests, 9, 10, 83, 93-94
Coot, 34
Copper butterfly, 32, 53
Corvidae (crow) family, 84, 85
Cottontail rabbit, 70, 73
Crickets, 77, 78
Crossbills, 94, 95, 97
Cryptic plumage, 34

Darter fish, 51
Deciduous forests, 9, 37-40, 83, 84, 96-98
Deer, 69-70, 94
Deerfly, 79, 81
Deer mice, 74
Diatoms, 23-24
Drumlin, 10
Ducks, 32-36
Dutch elm disease, 37

Eastern swift, 46
Eastern white pine, 83
Eft, tadpole, 44
Egrets, 18
Elders, 37

Fin whale, 100
Fireflies, 77-78
Fish
   of Great Lakes, 25, 28-29
   of northeast forest region, 50-51
Five-lined skink, 46
Flowering rush, 33
Flying squirrels, 70, 72-73
Forests
   amphibians in, 40
   coniferous, 9, 10-11, 83, 93-94
   deciduous, 9, 37-40, 83, 84, 96-98
   insects of, 79-81
   mammals of, 65-76
Fourhorn sculpin, 28
Freshwater coasts, 31-33
Frogs, 40-44

Gannets, 98, 101
Garter snake, 46-47
Glaciated valley, 11
Glaciers, 9, 10, 11-12, 18-19
Glossy ibis, 102, 103
Grasshoppers, 77, 78
Grasslands, 15
Gray fox, 75
Gray jay, 84-85, 94, 97
Gray squirrel, 72, 73
Great blue heron, 102, 103
Great Lakes
   area surrounding, 31-35
   described, 13, 23
   fish of, 25, 28-29
   pollution in, 7, 23-24, 31
   as route for explorers, 6
   swamps, 31, 33-35

Green frog, 41, 43
Green Mountains, 9
Ground squirrel, 71
Grouse, 84, 85, 93

Hellbender, 45, 46
Hibernation, 47, 50
Horned owl, 62, 63
Horseshoe crab, 104-105
Horseshoe Falls, 13
Hudson Bay, 11
Hudson River, 6
Humpback whale, 99-100

Indiana, areas of natural interest, 117
Indigo bunting, 61
Insects, 53-55, 79-81

Jays, 58-60, 84-85, 94, 97
Junco, 96

Kittiwake, 101-102

Lake Erie, 12, 13, 23, 30
Lake Huron, 28, 30
Lake Michigan, 30, 33
Lake Ontario, 12, 25, 30
Lake Superior, 22, 28, 30
Lake trout, 25, 28
Laomorpha order, 73
Lazuli bunting, 61
Leech, and frog eggs, 43
Lobster, 102
Long-billed rail, 102
Lynx, 89, 93, 94

Maine, areas of natural interest, 112
Malacostoma genus, 53
Malacostoma neustrium (tent caterpillar), 53
Mallard, 32
Mammals
  of forests, 65-76
  of northeast region, 85-90
  underground, 73-75
Maples, 37, 39
Marmots, 71
Maryland, areas of natural interest, 116-117
Massachusetts, areas of natural interest, 113
Meadow vole, 74
Metamorphosis, 41, 45-46
Michigan, areas of natural interest, 117, 120
Migration routes, birds, 60
Migratory birds, 57-58, 96
Minch River, 33
Moles, 73-74
Mollusk, 46
Moose, 88-89, 94
Moraine, 10, 19
Mosquitoes, 54

Moths, 53-55
Mountains
  Adirondacks, 36, 114-115
  Appalachian, 9
  Green, 9
  Rockies, 14
Mud dauber wasp, 78-79
Mud puppy, 46
Murres, 101
Muskie, 28
Muskrat, 88, 89
Mustelidae family, 93
Myrtle warbler, 61

New Hampshire, areas of natural interest, 112
New Jersey, areas of natural interest, 117
Newt, 44, 46
New York, areas of natural interest, 113, 116
Niagara Falls, 12, 13
Nonmigratory birds, 58-63
Northern flying squirrel, 70, 73, 90
Nuphar advena (yellow spatterdock), 33
Nuthatch, 63

Oaks, 37, 39
Opossum, 65-66, 75

Paper birch, 83
Paper wasp, 79
Pheromone, tent caterpillars, 55
Phytoplankton, 23-24
Pike, 29-30
Pine cones, 94-95
Pines, 83
Plains, 14-16
Planetree maple, 39
Plankton, 99
Plate tectonics, 9
Poison ivy and oak, 38
Poison sumac, 38-39
Pollution, 7, 20-21, 23-24
Polymorphism, 53
Porcupine, 86-87
Predators, 28, 34, 89

Quaking aspen, 83

Rabbit, 70, 73
Raccoon, 66
Rallidae family, 34
Rattlesnake, 47, 49
Raven, 85
Razorbills, 101
Red crossbill, 94, 95, 97
Red fox, 64
Red pine, 83
Red salamander, 45
Red-winged blackbird, 58
Reptiles, 46-50

Richardson's owl, 92, 93
Right whale, 101
Rocky Mountains, 14
Rodentia order, 73
Rodents, 70
Rosy finch, 61
Ruffed grouse, 60-61, 85, 93

Saint Lawrence River, 6, 13, 25, 26-27
Salamander, 44, 45, 46
Salmon, 25
Sandy beaches, 103
Seabirds, 101-103
Sea lamprey, 30
Sedimentary rock, 16
Seventeen-year locust, 78
Shagbark hickory, 38
Shiner fish, 51
Shorebirds, 103
Short-tailed shrew, 74
Shrews, 74-75, 92
Skimmer, 103, 104
Skunk, 67, 74
Skunk cabbage, 39
Small fauna, 52-56
Snakes, 46-49
Snapping turtle, 49-50
Snow belt, 13
Snowshoe hare, 85-86, 93
Solan goose, 101
Southern flying squirrel, 70, 72-73
Spathe, skunk cabbage, 39
Spike, 25
Spotted salamander, 45, 46
Spotted skunk, 67
Spring peeper, 41
Spruce, 83
Squirrels, 70, 72-73, 81, 90, 94
Staghorn sumac, 38
Starling, 57
Star-nosed mole, 73-74, 75
Sugar maple, 37, 39, 80-81
Swamps, 31, 33-35, 103
Sycamore, 37

Tadpoles, 41, 43, 44, 45
Tamarack, 83
Tent caterpillars, 53-55
Thrushes, 63
Tickseed-sunflowers, 76
Tidal swamps, 33
Timber rattlesnake, 47, 49
Titmice, 94, 97
Toad, 41
Toothed whales, 99
Tornadoes, 15-16
Tree frog, 41-42, 43
Tree swallow, 58, 110-111
Triassic period, 104
Turtles, 49-50

Vermont, areas of natural interest, 113

Verrucae, toads, 41
Virginia, areas of natural interest, 117
Voles, 74, 75, 92, 93

Wading birds, 103
Warblers, 61-63
Wasps, 78-79
Water lilies, 31
Water snake, 47, 48-49
Weasel, 92, 93
Welland Canal, 13, 30
Whales, 99
Whitefish, 25
White oak, 39
White spruce, 83
White-tailed deer, 69-70
White-throated sparrow, 97
White-winged crossbill, 97
Willow thrush, 63
Winter, 91-98
Wolf, 89, 95
Woodchuck, 70, 71
Woodcock, 57-58
Wood frog, 40
Wood thrush, 63

Yellow spatterdock, 33
Yellow warbler, 61-62

REFERENCE--NOT TO BE
TAKEN FROM THIS ROOM

	DATE DUE	

```
574.5    Wingfield, John C
Win
         Northeastern
         America
```

15.00

25597